A Student's Guide to Rotational Motion

Rotational motion is of fundamental importance in physics and engineering, and an essential topic for undergraduates to master. This accessible yet rigorous Student's Guide focuses on the underlying principles of rotational dynamics, providing the reader with an intuitive understanding of the physical concepts, and a firm grasp of the mathematics. Key concepts covered include torque, moment of inertia, angular momentum, work and energy, and the combination of translational and rotational motion. Each chapter presents one important aspect of the topic, with derivations and analysis of the fundamental equations supported by step-by-step examples and exercises demonstrating important applications. Much of the book is focused on scenarios in which point masses and rigid bodies rotate around fixed axes, while more advanced examples of rotational motion, including gyroscopic motion, are introduced in a final chapter.

EFFROSYNI SEITARIDOU is Professor of Physics at Oxford College of Emory University and received her Ph.D. in applied physics from Caltech in 2008. Her research is focused on biological physics. She has co-authored a book, *Simple Brownian Diffusion* (Oxford University Press, 2012). She has received numerous awards for excellence in teaching.

ALFRED C.K. FARRIS is Assistant Professor of Physics at Oxford College of Emory University and received his Ph.D. from the University of Georgia. He teaches the rotational dynamics module at Emory alongside Effrosyni Seitaridou, and has received awards for excellence in teaching both at Emory and at the University of Georgia.

Student's Guides

Other books in the Student's Guide series:

A Student's Guide to
Rotational Motion

EFFROSYNI SEITARIDOU
Emory University, Atlanta

ALFRED C.K. FARRIS
Emory University, Atlanta

CAMBRIDGE
UNIVERSITY PRESS

Shaftesbury Road, Cambridge CB2 8EA, United Kingdom

One Liberty Plaza, 20th Floor, New York, NY 10006, USA

477 Williamstown Road, Port Melbourne, VIC 3207, Australia

314–321, 3rd Floor, Plot 3, Splendor Forum, Jasola District Centre,
New Delhi – 110025, India

103 Penang Road, #05–06/07, Visioncrest Commercial, Singapore 238467

Cambridge University Press is part of Cambridge University Press & Assessment,
a department of the University of Cambridge.

We share the University's mission to contribute to society through the pursuit of
education, learning and research at the highest international levels of excellence.

www.cambridge.org
Information on this title: www.cambridge.org/highereducation/isbn/9781009213349

DOI: 10.1017/9781009213349

First published 2023

A catalogue record for this publication is available from the British Library.

Library of Congress Cataloging-in-Publication Data
Names: Seitaridou, Effrosyni, author. I Farris, Alfred C. K., 1990– author.
Title: A student's guide to rotational motion / Effrosyni Seitaridou, Emory
University, Atlanta, Alfred C.K. Farris, Emory University, Atlanta.
Description: Cambridge, United Kingdom ; New York, NY : Cambridge
University Press, 2023. I Series: Student's guide I Includes
bibliographical references and index.
Identifiers: LCCN 2022062274 I ISBN 9781009213301 (hardback)
I ISBN 9781009213356 (paperback) I ISBN 9781009213349 (ebook)
Subjects: LCSH: Rotational motion – Textbooks. I Rotational
motion (Rigid dynamics) – Textbooks.
Classification: LCC QC133 .S45 2023 I DDC 531/.34–dc23/eng20230512
LC record available at https://lccn.loc.gov/2022062274

ISBN 978-1-009-21330-1 Hardback
ISBN 978-1-009-21335-6 Paperback

Additional resources for this publication at www.cambridge.org/seitaridou-farris

About this book

This edition of *A Student's Guide to Rotational Motion* is supported by an extensive range of interactive digital resources, available via a companion website. These resources have been designed to support your learning and bring the textbook to life, supporting active learning and providing you with feedback. Please visit www.cambridge.org/seitaridou-farris to access this extra content.

We may update our Site from time to time, and may change or remove the content at any time. We do not guarantee that our Site, or any part of it, will always be available or be uninterrupted or error free. Access to our Site is permitted on a temporary and "as is" basis. We may suspend or change all or any part of our Site without notice. We will not be liable to you if for any reason our Site or the content is unavailable at any time, or for any period.

Contents

Preface

Rotational motion is ubiquitous in nature, as well as in the undergraduate- and graduate-level physics curriculum. Unfortunately, the more complex mathematics, the lack of time to do justice to this unit in introductory courses, and the lack of intuition on the topic often lead to confusion and misconceptions. Our goal here is not to cover everything about rotational motion, but rather to present the simplest case of this motion thoroughly, both in the mathematical and in the physical sense. Namely, we will focus on the motion of point masses and extended rigid bodies about *fixed axes* (meaning an axis that does not change its orientation over time) from the point of view of an *inertial observer*.

But why opt for depth at the expense of breadth? One of our goals for writing this book is to showcase scientific thinking to students who are not yet experienced scientists – to convey how scientists think when they approach new problems. Their thought process is methodical and draws parallels from existing knowledge; these are the features we try to highlight in our treatment of the simplest cases of rotational motion. In our view, it is not the breadth of rotational motion or the abstract mathematical formalism one can use to study it that makes this unit so enticing to write about. Rather, rotational motion is enticing because even its simplest problems are complex enough to demonstrate patterns, both in the laws of nature and in the way humans approach nature's greatest mysteries.

In writing this book, we are assuming that our readers have had calculus and understand the basic ideas of scalar and vector products. We are also assuming that our readers have a good understanding of kinematics, Newton's laws of motion, and energy and momentum theorems as they apply to linear motion. Our approach is to extend these ideas to rotational motion to showcase the similarities and thus how scientists approach new problems. To better communicate with our audience, we have adopted a casual, conversational style but have remained thorough and rigorous in our description and analysis of the

ix

presented ideas. To help us achieve our goals, we have structured each chapter so that it includes three components: (a) The theory and presentation of the chapter's concepts with a focus on drawing analogies from linear motion, for which the students have a better developed intuition and understanding. We are also intentional about addressing subtleties that are often overlooked or superficially explained. (b) The derivations and detailed analysis of fundamental equations in rotational motion. (c) A detailed presentation of solutions to thoughtful examples that illustrate the appropriate application of the concepts and systematic problem-solving methodology that can be applied throughout one's career as a physicist.

We end this preface with a piece of heartfelt advice. Some sections of this book might be difficult to grasp initially. However, as we often say to our students, struggling to understand difficult concepts allows us to become resilient and grow in the process. The key is to keep at it. When you find yourself on the other side – when you have understood a difficult concept – not only will you feel accomplished, but you will also feel more prepared to approach increasingly complex problems. This willingness to approach new problems is a key trait of a good scientist and, as teachers, the greatest joy for us is the opportunity to train good scientists.

1
Rotational Kinematics

Communication in any language requires the proper use of its vocabulary. Thus, to clearly communicate physical phenomena, we first have to learn our physics vocabulary. Since this is the first chapter, let's start with some preliminary definitions of physical quantities that will show up throughout our study of rotational motion.

1.1 Preliminary Definitions

- **Periodic motion:** A motion that repeats itself (in the same way) at equal time intervals. For example, each hand of a clock executes periodic motion.
- **Period (T):** The time required to complete one repetition of a periodic motion. The SI unit is the second [s]. For example, the period of the hour hand of the clock is 12 hours = 12 hours · 3,600 s/hour = 43,200 s.
- **Frequency (f):** The fraction given by the number of repetitions N of the periodic motion over a time interval Δt [s] divided by the time interval Δt:

$$f = \frac{N}{\Delta t}. \tag{1.1}$$

Equivalently, if $\Delta t = 1$ s, then frequency is the number of repetitions within 1 s. The SI unit of frequency is the Hertz [Hz]: 1 repetition/s = 1 Hz. For example, the frequency of the hour hand of the clock is $\frac{1}{43,200}$ Hz because it executes 1 revolution in 43,200 s (12 hours).

From the definition of the period and the frequency, we see that if the time interval Δt is equal to the period T, then the number of repetitions N is equal to 1. Therefore,

$$f = \frac{1}{T} \Rightarrow T = \frac{1}{f}. \tag{1.2}$$

- **Rigid body:** A solid object in which the intermolecular forces are so strong that when external forces are exerted, the deformation is small enough to be neglected. For a rigid body, therefore, the distance between any two points of the object will not change over time. Our focus in this book will be on the rotational motion of rigid bodies.

- **Rotational motion about a fixed axis of rotation:** This is a type of motion in which an object moves about an axis that does not change its orientation (direction) in space. Even if the axis of rotation moves linearly, it is considered fixed with respect to rotational motion as long as its orientation is not changing. There are two types of rotational motion: (1) orbital rotational motion and (2) spin rotational motion. It is possible for an object to exhibit both orbital and spin rotational motion as is the case of the Earth, for example, which executes orbital motion around the sun while also spinning about its own axis every 24 hours, on average. Rotational motion can be periodic but does not have to be. Rotational motion can also be executed by point masses, which we assume have infinitely small dimensions, as well as extended solid bodies whose dimensions cannot be neglected.

As we will see shortly, rotational motion cannot be quantified without specifying a rotation axis – the axis of rotation provides a reference for the physical quantities of rotational motion, both for extended rigid bodies and point masses. Just like in translational motion where one can pick any coordinate system with respect to which the motion can be analyzed, in rotational motion, one can pick any axis of rotation. However, once we choose an axis of rotation, we must stick to that choice for our calculations to be consistent. In this book, when an object is executing spin rotational motion, we will choose for the axis of rotation the straight line through all fixed points of the body around which all other points of the body move in circles (i.e., rotate around). Why? Because this will greatly simplify our calculations. As an example of this choice of axis of rotation, think about a spinning disk that is constrained to rotate about its center as shown in Figure 1.1a (e.g., a vinyl record rotating on a record player). As this disk is spinning, the center of the disk is fixed, while all the other points of the disk move in circles. Thus, the axis of rotation we choose in this case, as shown in Figure 1.1a, will be the line that goes through the center of the disk and is perpendicular to the plane of the disk.

The axis of rotation does not have to go through points that belong to the object, as is often the case in orbital motion. When an object is executing orbital rotational motion, we will choose for the axis of rotation the straight line about which the body moves in circles. For example, as shown in Figure 1.1b, during its orbital motion, a mass can rotate about an axis that is outside of the object. For the small ball shown, the axis of rotation is a rod that connects to the ball

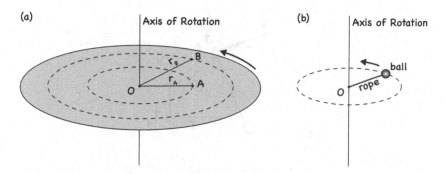

Figure 1.1 (a) For a spinning disk pinned at its center O, the axis of rotation goes through the disk's center and is perpendicular to its plane. The center of the disk remains fixed while all other points (e.g., points A and B) traverse circular trajectories. (b) The small ball is tethered to a rope whose other end O is fixed. The ball is executing orbital motion, traveling in a circle about an axis that goes through point O and is perpendicular to the circular trajectory. In this case, the axis of rotation does not go through the object.

via the rope and is in the center of the circle the ball traverses. One can see that if the rotating object is a point mass, then it will behave like the ball of Figure 1.1b.

As mentioned in the preface, in this book, we focus on rotation about a *fixed* axis from the viewpoint of an inertial observer, which is the simplest kind of rotational motion. So why are we limiting our study to only the simplest case? This is because all the quantities we study in this book are fundamental to understanding the more general cases and more mathematically abstract formalism you will encounter in upper-level coursework. In Chapter 8, we will briefly study a case where the axis of rotation is not fixed and present some of the mathematical formalism used to describe rotations in upper-level mechanics courses.

1.2 Rotational Motion with Constant Speed

With these definitions in hand, we can already start studying some physical phenomena. Recall from our course in introductory mechanics, one of the first concepts encountered was kinematics with constant (translational) velocity and constant (translational) acceleration. So, let's start with the simplest, equivalent phenomena for rotational motion.

First, let's consider a point mass that moves in a circle and travels equal arc lengths at equal time intervals (i.e., the speed remains constant). In this motion, the linear velocity $\vec{v}$ has constant magnitude, but its direction changes.

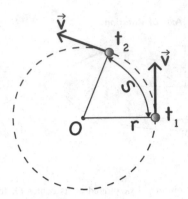

Figure 1.2 Particle rotating with constant speed along a circular trajectory of radius r about an axis that goes through the center of the circle O and is perpendicular to the page. The velocity vector $\vec{v}$ changes direction as the object traverses arc length S over the time interval $\Delta t = t_2 - t_1$.

The velocity vector is always tangent to the trajectory (in this case the trajectory is the circle), and it points in the direction of motion. Figure 1.2 shows the velocity vector at two different time instants (t_1 and t_2, with $t_2 > t_1$) as the point mass moves around the circle in the counterclockwise direction with constant speed.

Since the particle is moving with constant speed, its average speed is the same as its instantaneous speed. If the particle travels an arc length S in a time interval Δt, then the speed (average or instantaneous) will be the arc length divided by the time interval:

$$v = \frac{S}{\Delta t} = \frac{S}{t_2 - t_1}. \tag{1.3}$$

Let's explore some additional relations. During one period T, by definition, the particle will have completed one revolution. Therefore, it will have traveled an arc length equal to the circle's circumference $2\pi r$, where r is the radius of the circle. Thus, we have:

$$v = \frac{S}{\Delta t} = \frac{2\pi r}{T} = 2\pi r f, \tag{1.4}$$

where in the last relation we made use of Equation (1.2).

Let's look more carefully at the implications of Equation (1.4). It is important to point out that while we have derived this equation in the context of a point mass, it is readily applicable to extended, rigid objects in the case where all points of the rigid body execute circular motion about the axis of rotation. As an example, let's look again at a disk rotating with constant speed about an axis

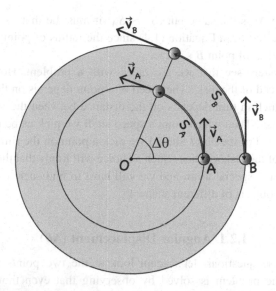

Figure 1.3 Two points on a disk, A and B, rotating about the same axis of rotation perpendicular to the page and through point O. The two points are at different distances from the axis of rotation, and each traverses a different arc length, S_A and S_B, respectively, in the same time interval Δt. Thus, their respective speeds v_A and v_B are different. However, during this time interval they both traverse the same angle $\Delta\theta$.

perpendicular to its plane, going through its center, as shown in a top-down view in Figure 1.3. Since this is a rigid body, all points must rotate with the same frequency; otherwise, the disk would deform. Therefore, Equation (1.4) shows us that the speed v of a particular point on the disk depends only on the radius r of the circle it transcribes (i.e., its distance from the axis of rotation).

With this in mind, now let's look at three points on the disk, namely points O, A, and B of Figure 1.3. Since these points are located at different distances from the axis of rotation that goes through the center of the disk, they will not have the same speed v because the radii of the circles they are transcribing are different. For point O, $r = 0$ and thus $v = 0$ according to Equation (1.4). The two points A and B, located along the same radial line, will traverse different arc lengths during the same time interval as the disk rotates by an angle $\Delta\theta$. For the two points, respectively, we have that:

$$v_A = \frac{S_A}{\Delta t},$$

$$v_B = \frac{S_B}{\Delta t}.$$

(1.5)

However, since Δt is the same but $S_B > S_A$, it must be that $v_B > v_A$. This result was expected from Equation (1.4) since the radius of point A's circle is less than the radius of point B's circle.

We immediately see that we are faced with a problem. How does one describe the speed of the disk? The speed of a point depends on the arc length it traverses, which in turn depends on the distance between the point and the center of the disk. Then which point's speed shall we pick as being representative of the speed of the disk? Should we pick a point on the rim of the disk? The midpoint of the disk's radius? Such a choice will imply that there is a point on the disk that is special to us, and we will have to find such a special point for all rotating objects of different shapes!

1.2.1 Angular Displacement ($\Delta\theta$)

To answer these questions, let's again look at the two points A and B of Figure 1.3. The problem is solved by observing that even though the two points traverse different arc lengths, they both traverse the same angle $\Delta\theta$. We therefore conclude that this angle represents a better way of describing the disk's displacement during rotation than does the traversed arc length, which is different for points at different distances from the center.

Now let's examine Figure 1.4. In the figure, θ by itself represents the angular position for the points on the disk, conventionally measured with respect to the $+x$ axis and regarded as positive when the rotation is in the counterclockwise direction. Thus, all points that belong to the same radial line (e.g., OP or OP′) will have the same angular position with respect to the $+x$ axis (θ_1 or θ_2, respectively).

In Figure 1.4, if we follow the motion of any point on the disk that is initially along the radius OP, such as point A, we see that this point has an angular position θ_1 at time t_1, and an angular position θ_2 at time t_2, when it lies along the radial line OP′. Therefore, its angular displacement within that time interval is defined by:

$$\Delta\theta = \theta_2 - \theta_1 = \theta_{\text{final}} - \theta_{\text{initial}}. \tag{1.6}$$

If the disk is rigid, then all its points will have the same angular displacement during the same time interval, even though they did not travel equal arc lengths. If they did not have the same angular displacement, then the distances between two such points would change over time and thus the disk would deform. Since the angular displacement is the same for all points on the disk, it is more convenient to use than the arc length when studying the displacement of this rotating object as a whole. In summary, just like the linear displacement $\Delta\vec{x}$ is defined

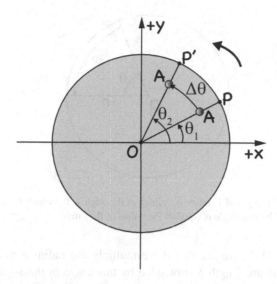

Figure 1.4 As the disk rotates counterclockwise, point A, which was initially on line OP at time t_1, moves to line OP' at time t_2. Therefore, its angular position changes from θ_1 at time t_1 to θ_2 at time t_2. Its angular displacement $\Delta\theta$ is given by Equation (1.6). Since the disk is rigid, all its points will have the same angular displacement as point A during the same time interval.

as the change in the position $\vec{x}$ of an object, the angular displacement $\Delta\theta$ is the change in the angular position θ of any point of the object and, thus, of the whole object.

Units of Angular Displacement

At this point, you may be wondering whether the angular position and displacement are measured in radians, revolutions, or degrees. While, in principle, one could express the angular position or displacement in degrees or revolutions, the radian is the most natural choice. To see why, let's examine Figure 1.5. An angle of 1 radian [rad] is defined as the angle θ for which the arc length S subtended by this angle is equal to the radius of the circle r. In general, the angle θ (in radians) is given by the ratio:

$$\theta = \frac{S}{r} \Rightarrow S = r\theta, \tag{1.7}$$

so that if $S = r$, then $\theta = 1$ rad.

The reason why the radian is a natural way of measuring angles is because it is, by definition, the ratio of two physical quantities with the same units, in contrast to the degree for which there are (arbitrarily) 360 in a circle. Thus,

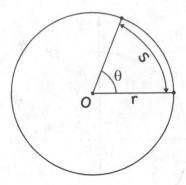

Figure 1.5 An angle of 1 radian is defined as the angle θ for which the arc length
S subtended by this angle is equal to the radius of the circle r.

from Equation (1.7), we see that if we multiply the radius r by some angle
θ, we obtain the arc length S subtended by this angle *in the same units as* r,
provided that θ is in radians. The radian is preferable because of the natural
link it provides between angular and linear quantities. If we need to convert
between units of measurement, 1 revolution $= 2\pi$ rad $= 360°$. Looking back at
Figure 1.3, we see that in radians, the angular displacement $\Delta\theta$ of points A and
B is given by:

$$\Delta\theta = \frac{S_A}{OA} = \frac{S_B}{OB}. \tag{1.8}$$

1.2.2 Angular Velocity ($\vec{\omega}$)

Following our analogy between $\Delta\theta$ and $\Delta\vec{x}$ from translational motion, we now
define the concept of velocity for rotational motion. The magnitude of the **average angular velocity** ω_{avg} is an object's angular displacement $\Delta\theta$ that occurs
within a time interval Δt divided by this time interval Δt:

$$\omega_{avg} = \frac{\Delta\theta}{\Delta t} = \frac{\theta_{final} - \theta_{initial}}{t_{final} - t_{initial}}. \tag{1.9}$$

Angular velocity is actually a vector quantity (which we will discuss in more
detail shortly) with the SI unit of radians per second [rad/s] and direction
pointing along the axis of rotation of the object. For the case of the rotating
disk fixed at its center O shown in Figure 1.6, the axis of rotation is perpendicular to the plane of the disk and goes through the center O, as defined earlier.
$\vec{\omega}_{avg}$ would then be along this axis. But how do we get the correct direction
(i.e., up or down) along the axis of rotation?

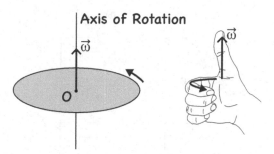

Figure 1.6 If the disk is rotating counterclockwise as seen from above, then the right-hand rule dictates that the angular velocity will be pointing upwards as shown. This is because if we grasp the axis of rotation with our right hand so that our four fingers curl in the direction of rotation of the object, then our extended thumb points in the direction of the angular velocity vector. The angular velocity vector is drawn along the axis of rotation. If the angular velocity is constant, then the instantaneous angular velocity $\vec{\omega}$ is equal to the average angular velocity $\vec{\omega}_{avg}$.

The direction at any moment is given by the right-hand rule. The right-hand rule says that if we grasp the axis of rotation with our right hand so that our four fingers curl in the direction of rotation of the object, then our extended thumb points in the direction of the angular velocity vector, as shown in Figure 1.6. This version of the right-hand rule is actually a shortcut; we will learn the more general right-hand rule soon when we talk about the more mathematically rigorous relation between linear velocity and angular velocity.

By definition, we also see that if Δt is equal to 1 period (i.e., $\Delta t = T$), then $\Delta \theta$ will be equal to 2π [rad]. Thus,

$$\omega_{avg} = \frac{\Delta \theta}{\Delta t} = \frac{2\pi}{T} = 2\pi f, \tag{1.10}$$

where the last relation came from the use of Equation (1.2).

Why Is Angular Velocity a Vector?

Now we have to open a subtle but important parenthesis: Based on the definition of ω_{avg} given in Equation (1.9), since Δt is a positive scalar ($t_{final} > t_{initial}$), $\Delta \theta$ has to be a vector for ω_{avg} to be a vector. This implies that θ has to be a vector (from Equation (1.6)). Moreover, since Δt is a positive scalar, from Equation (1.9) we see that $\vec{\omega}_{avg}$ must have the same direction as the vector $\Delta\vec{\theta}$.

If we define $\vec{\theta}$ as a vector whose magnitude is the angular position θ and whose direction runs parallel to the axis of rotation, with the direction of $\Delta\vec{\theta}$ also given by the right-hand rule, then this definition certainly resolves the problem with the definition of $\vec{\omega}_{avg}$. But is this all we need? Can we define the vector quantity $\vec{\theta}$?

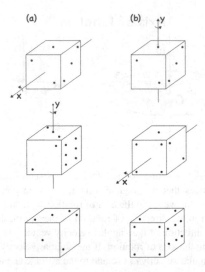

Figure 1.7 A die is rotated about two different axes, x and y, by the same amount of 90°. In case (a), we rotate about the x axis first. In case (b), we rotate about the y axis first. The final states of the die are not the same, indicating that the two rotations do not commute.

Just because a quantity has a defined magnitude and direction does not mean it is a vector. Vectors obey the commutative property. That is, for two vectors $\vec{a}$ and $\vec{b}$:

$$\vec{a} + \vec{b} = \vec{b} + \vec{a}. \tag{1.11}$$

However, this commutative property does not hold for rotation in general. For example, as shown in Figure 1.7, the same set of rotations of a die about two different axes does not result in the same outcome if the rotations in the set are performed in the reverse order. However, if the rotation angles are small (i.e., infinitesimal), the rotation is almost planar, just like the rotation of the disk about the axis shown in Figure 1.6. As we know, planar displacements commute; for example, there is no difference in a particle's final location if it moves 2 m in the $+y$ direction and then 1 m in the $+x$ direction versus if it moves 1 m in the $+x$ direction and then 2 m in the $+y$ direction. Thus, in this case, the two (infinitesimal) angular displacement vectors will commute with each other and $d\vec{\theta}_1 + d\vec{\theta}_2 = d\vec{\theta}_2 + d\vec{\theta}_1$. Therefore, though θ is not a vector, $d\vec{\theta}$ is! As a final note, while rotations do not commute in general, two rotations about the same axis of rotation do obey the commutative property because they are planar.

Knowing that $d\vec{\theta}$ is a vector whose magnitude is the angle of rotation ($d\theta$) and whose direction is given by the right-hand rule, as described above, is immensely convenient because it allows us to define the concept of the **instan-taneous angular velocity** $\vec{\omega}$ in a way that does justice to what we know about vectors and maintains the analogy with translational motion. If an object rotates by an infinitesimally small angle $d\vec{\theta}$ during an infinitesimally small time interval dt, then

$$\vec{\omega} = \frac{d\vec{\theta}}{dt}. \tag{1.12}$$

As a result, $\vec{\omega}$ is always a vector, and it obeys the commutative property. $\vec{\omega}$ is the instantaneous rate of change of angular position θ just like $\vec{v}$ is the instantaneous rate of change of position $\vec{x}$ in the case of linear motion. If the points of a rigid object execute circular motion about the axis of rotation, $d\vec{\theta}$ is the same for all points, and thus so is $\vec{\omega}$. Unlike the linear velocity of a point on the disk $\vec{v}$, which depends on r (based on Equations (1.4) and (1.5)), $\vec{\omega}$ is the same for all points on the rotating object regardless of r. Therefore, $\vec{\omega}$ is the representative velocity for the rotating object, just like $\Delta\theta$ was its representative displacement!

Finally, we understand that if the instantaneous angular velocity $\vec{\omega}$ is constant, then it will be equal to the average angular velocity $\vec{\omega}_{\text{avg}}$. In that case (for magnitudes),

$$\omega_{\text{avg}} = \omega$$
$$\Rightarrow \frac{\Delta\theta}{\Delta t} = \frac{d\theta}{dt}, \tag{1.13}$$

and Equation (1.10) will apply for the instantaneous angular velocity also. If the instantaneous angular velocity is not constant, then the frequency and the period will not be constant either. In this scenario, Equation (1.10) can also be used under the condition that we use the instantaneous period and frequency of rotation.

One might ask: Why pick the direction of $\vec{\omega}$ (or $d\vec{\theta}$) to be along the axis of rotation? This is another subtlety: The simple rotational motion we are studying here is a motion in two dimensions because any rotating point is executing circles and, thus, is moving in a plane. The orientation of a plane of rotation is described by its normal vector. So, when an object is rotating in a plane, we can use a vector normal to the plane to indicate that motion. In addition, we pick that normal vector to be along the axis of rotation, so as to indicate about which axis the angular position changes and in which direction. In other words, having the angular velocity be perpendicular to the plane of motion along the axis of rotation allows us to both describe the orientation of the plane of motion and also to define the axis around which the angle changes.

There is an additional benefit to representing a planar rotation with a vector that is along the axis of rotation: It allows us to reduce this two-dimensional

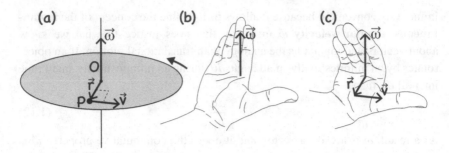

Figure 1.8 (a) The linear velocity of any rotating point, such as point P of a rotating disk, can be found via the cross product of the angular velocity vector $\vec{\omega}$ and the position vector $\vec{r}$ that goes from the axis of rotation to the point. (b) To find the direction of $\vec{v}$ using the right-hand rule, we first need to orient the four fingers of our right hand to point in the direction of the first vector in the cross product ($\vec{\omega}$). (c) Then, we bend our fingers to align them with the second vector in the cross product ($\vec{r}$) and our extended thumb points in the direction of the resulting vector ($\vec{v}$).

motion, which would be more difficult to study, into a one-dimensional problem, where quantities are defined along the one-dimensional axis of rotation. This reduction from two dimensions to one allows us to employ all the principles we learned when studying one-dimensional translational motion. In other words, as we saw in Figure 1.2, although $\vec{v}$ is always changing directions in the plane of the object's rotation, $\vec{\omega}$ is always along one axis (i.e., the axis of rotation). We will see this throughout our study of rotational motion – the physical quantities we will use to describe rotational motion in this book, if vectors, will have their direction along the axis of rotation.

Angular Velocity: Cross Products and the Right-Hand Rule

Let's now be a little more mathematically rigorous. The fact that we have to use the right-hand rule when finding the direction of $\vec{\omega}$ implies that there is a cross product of two vectors somewhere. Since we can describe the speed of an object undergoing rotational motion using either the angular velocity or *both* the linear velocity and the radius of the circle this object traverses, it is reasonable to assume there is a conversion between the velocities. Let's examine a point on a rotating disk, such as point P in Figure 1.8a, and define the position vector $\vec{r}$ to be the vector that goes from the axis of rotation to the point. Then the magnitude of $\vec{r}$ is the radius of the circle that this point transcribes. The angular velocity vector can be defined as:

$$\vec{v} = \vec{\omega} \times \vec{r}. \tag{1.14}$$

Here we have the cross product of the two vectors $\vec{\omega}$ and $\vec{r}$. To find the direction of the resulting vector $\vec{v}$, we use the version of the right-hand rule where we place the four fingers of our right hand in the direction of the first vector in the cross product ($\vec{\omega}$) (Figure 1.8b) in an orientation so that when we bend our fingers to align them with the second vector in the cross product ($\vec{r}$), our extended thumb points in the direction of the linear velocity $\vec{v}$ (Figure 1.8c). Because $\vec{v}$ is the result of a cross product, $\vec{v}$ is perpendicular to both $\vec{\omega}$ and $\vec{r}$ individually, as well as to the plane formed by $\vec{\omega}$ and $\vec{r}$.

In addition, we can see that if we substitute the definitions for $\vec{v} = \frac{d\vec{S}}{dt}$ and $\vec{\omega} = \frac{d\vec{\theta}}{dt}$ in Equation (1.14) and cancel dt from both sides of the equation, we get:

$$\vec{v} = \vec{\omega} \times \vec{r}$$

$$\Rightarrow \frac{d\vec{S}}{dt} = \frac{d\vec{\theta}}{dt} \times \vec{r}$$

$$\Rightarrow d\vec{S} = d\vec{\theta} \times \vec{r}. \tag{1.15}$$

Here we obtain a cross product and right-hand rule similar to what we saw in Equation (1.14) since $d\vec{S}$ has the same direction as $\vec{v}$ and $d\vec{\theta}$ has the same direction as $\vec{\omega}$.

Since we are looking at rotations about fixed axes, $\vec{\omega}$ is perpendicular to $\vec{r}$, and we see from Equation (1.14) that for the magnitude v, we get:

$$v = \omega r \sin \phi = \omega r \sin \left(\frac{\pi}{2}\right) = \omega r. \tag{1.16}$$

In a similar way, $d\vec{\theta}$ is perpendicular to $\vec{r}$. Then, starting from Equation (1.15), we get for the magnitude ds:

$$ds = r d\theta \sin \phi = r d\theta \sin \left(\frac{\pi}{2}\right) = r d\theta. \tag{1.17}$$

Lastly, it is worth noting that Equation (1.16) can also be obtained by combining the relations $v = \frac{2\pi r}{T}$ and $\omega = \frac{2\pi}{T}$. Now, let's do an example that will help us clarify some of these definitions.

Example 1: Orbital rotational motion with constant speed. A small object is moving clockwise (as seen from above) around a circle of radius $r = 20$ m with constant speed. The object has traveled around the circle 100 times within time $\Delta t = 20$ s. Find (a) the frequency and the period of the object's motion and (b) the angular velocity of the motion. (c) Draw a diagram showing the object's angular and linear velocities for an arbitrary position of the object along the circle.

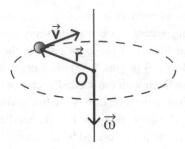

Figure 1.9 The object is executing clockwise (as seen from above) orbital circular motion. The right-hand rule shows that the angular velocity is pointing downward along the axis of rotation. The cross product in Equation (1.14) can be used to find the direction of the linear velocity vector at this arbitrarily chosen location for the object.

(a) Based on the definitions for frequency (Equation (1.1)) and period (Equation (1.2)) we have:

$$f = \frac{N}{\Delta t} = \frac{100 \text{ rev}}{20 \text{ s}} = 5.0 \text{ Hz} \tag{1.18}$$

and

$$T = \frac{1}{f} = \frac{1}{5.0 \text{ s}^{-1}} = 0.2 \text{ s}. \tag{1.19}$$

(b) Since we know from Equation (1.10) that $\omega = 2\pi f$, then

$$\omega = 2\pi f = 2\pi \text{ rad} \cdot 5 \text{ s}^{-1} = 10\pi \text{ rad/s}. \tag{1.20}$$

(c) The diagram is shown in Figure 1.9. The axis of rotation passes through the center of the circle that the object traverses and is perpendicular to the plane of its motion. We must use the right-hand rule to find the direction of $\vec{\omega}$. First, we grasp the axis of rotation with our right hand. Since the object is moving clockwise around the circle, we curl the four fingers of our right hand clockwise along the trajectory of the object in the plane of the circle. Now, our extended thumb points downward, which therefore indicates that the direction of $\vec{\omega}$ is downward along the axis of rotation. The direction of $\vec{v}$ can be found by using the cross product in Equation (1.14) or the fact that the linear velocity vector is tangent to the trajectory of the circular motion of the object pointing in the direction of the motion.

1.3 Acceleration in Rotational Motion

1.3.1 Centripetal Acceleration ($\vec{a}_c$)

Imagine that we have a point mass moving around a circle with constant linear speed (i.e., constant angular velocity). The object's linear velocity is tangent to the circle while its angular velocity is along the axis of rotation in the direction consistent with the right-hand rule. As shown in Figure 1.10a, if the mass is rotating counterclockwise (as viewed from above), then $\vec{\omega}$ is pointing out of the page.

In this case, there is no linear acceleration (tangent to the circle), since the object's speed is not changing. However, the linear velocity *vector* is changing! Even though its magnitude is constant, its direction changes as the object traverses the circle.

Average acceleration is defined as

$$\vec{a}_{\text{avg}} = \frac{\Delta\vec{v}}{\Delta t}. \tag{1.21}$$

Let's assume that at time t_i the particle is at point A and has velocity $\vec{v}_{\text{initial}} = \vec{v}_A$ while at time t_f the particle is at point B and has velocity $\vec{v}_{\text{final}} = \vec{v}_B$. Then if we perform the vector subtraction

$$\Delta\vec{v} = \vec{v}_{\text{final}} - \vec{v}_{\text{initial}} = \vec{v}_B - \vec{v}_A, \tag{1.22}$$

as shown in Figure 1.10b, we will find that the resulting vector $\Delta\vec{v}$ has a direction that is pointing toward the center of the circle. Since Δt is a positive scalar, Equation (1.21) shows that the acceleration will also be pointing towards the

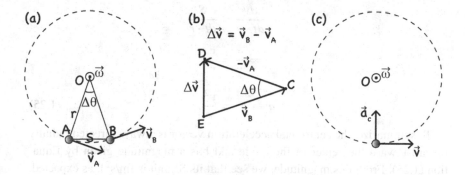

Figure 1.10 A mass moving around the circle with constant speed (a) has a non-zero change in velocity, $\Delta\vec{v}$, that points towards the circle's center (b). Thus, it has an acceleration pointing in the same direction (c), called centripetal acceleration.

center of the circle (Figure 1.10c). For this reason, this acceleration is called
centripetal (i.e., center seeking) acceleration. If we take the limit as Δt goes to
zero in Equation (1.21) to obtain the instantaneous acceleration, then points A
and B in Figure 1.10a will coincide in the middle of the arc length S. This is
why the centripetal acceleration in Figure 1.10c is drawn in the middle of the
arc length.

Naturally, if the velocity's direction is changing at a constant rate (i.e., by
the same angle at equal time intervals), the magnitude of the average centripetal
acceleration will be the same as the magnitude of the instantaneous centripetal acceleration. But what is this magnitude? For a derivation let's look at
Figure 1.10: Isosceles triangle OAB formed by the radii and the chord AB is
similar to isosceles triangle CDE, which is formed by the velocity vectors at
points A $(\vec{v}_A)$ and B $(\vec{v}_B)$ and the vector $\Delta\vec{v}$. This is because $\vec{v}_A$ and $\vec{v}_B$ are perpendicular to OA and OB, respectively, since the velocities are tangent to the
circle. Using the properties of similar triangles, we now have ($v_A = v_B = v$):

$$\frac{\Delta v}{v} = \frac{AB}{r}. \tag{1.23}$$

At the limit as Δt goes to zero (i.e., Δt becomes dt), the arc length S and the
chord length AB are equal (i.e., $\lim_{\Delta t \to 0} AB = S$). In addition, Δv becomes dv,
and Equation (1.23) becomes:

$$\frac{dv}{v} = \frac{S}{r}. \tag{1.24}$$

Dividing both sides of Equation (1.24) by dt and using the relations $a = \frac{dv}{dt}$
and $v = \frac{S}{dt}$, we obtain the magnitude of the centripetal acceleration a_c:

$$\frac{dv}{dt \cdot v} = \frac{S}{dt \cdot r}$$

$$\Rightarrow \frac{a}{v} = \frac{v}{r}$$

$$\Rightarrow a \to a_c = \frac{v^2}{r}. \tag{1.25}$$

To summarize, the centripetal acceleration vector is always directed radially
inward toward the center of the circle and has a magnitude given by Equation (1.25). From this magnitude, we see that its SI unit is [m/s^2], as expected.
If an object is moving in a circular trajectory, the centripetal acceleration
is *always* present, even if the mass is moving at constant speed. The centripetal acceleration is due to the change in the *direction* of the linear velocity
vector.

Centripetal Force

But what causes the centripetal acceleration? In other words, what causes the direction of the velocity vector to change? Let's remind ourselves of Newton's Second Law as we learned it for translational motion:

$$\Sigma \vec{F} = m\vec{a} \Rightarrow \vec{a} = \frac{\Sigma \vec{F}}{m}. \tag{1.26}$$

Equation (1.26) says that a net force $\Sigma \vec{F}$ causes a mass m to accelerate with an acceleration of $\vec{a}$. The only way we can have acceleration is to have a nonzero net force. The mass m expresses how difficult it is to get an object to accelerate (i.e., change its velocity). Given that m is a positive scalar quantity, the direction of the acceleration $\vec{a}$ will be that of the net force $\Sigma \vec{F}$.

As a result, to have a centripetal acceleration, the net force exerted on the mass must have a component that points toward the center of its circular trajectory. This component is called the centripetal force $\Sigma \vec{F_c}$ and its magnitude is given by combining Equation (1.26) with the result of Equation (1.25):

$$\Sigma F_c = ma_c = m\frac{v^2}{r}. \tag{1.27}$$

The centripetal force is not a different type of force (like the gravitational force or the force of friction, etc.), but rather refers to a direction: the centripetal force is the component of the *net* force that points toward the center of the object's circular trajectory. We get the centripetal force by adding all the force components that point toward and away from the center of the circle (i.e., are along the radial direction). Instead of calling it "net radial" we just call it "centripetal," since this adjective also specifies the direction.

If there is also a component of the net force on the object in the direction tangent to the circle, then, based on Newton's Second Law, this component will provide a tangential acceleration $\vec{a_t}$. The tangential acceleration would cause the mass to either speed up or slow down. This brings us to the next section.

1.3.2 Angular Acceleration ($\vec{\alpha}$)

Once again, we will follow our analogy from translational motion to define the concept of acceleration for rotational motion.

We define **average angular acceleration** $\vec{\alpha}_{\text{avg}}$ to be the change in an object's angular velocity $\Delta\vec{\omega}$ over a time interval Δt divided by the time interval Δt. If at an initial time t_{initial} the object's angular velocity is $\vec{\omega}_{\text{initial}}$ and at a final time t_{final} it is $\vec{\omega}_{\text{final}}$, then

$$\vec{\alpha}_{\text{avg}} = \frac{\vec{\omega}_{\text{final}} - \vec{\omega}_{\text{initial}}}{t_{\text{final}} - t_{\text{initial}}} = \frac{\Delta\vec{\omega}}{\Delta t}. \tag{1.28}$$

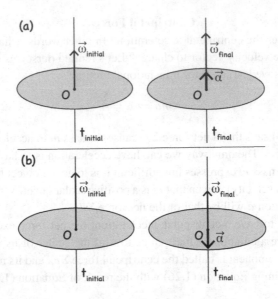

Figure 1.11 At an initial time t_{initial} an object's angular velocity is $\vec{\omega}_{\text{initial}}$ and at a final time t_{final} it is $\vec{\omega}_{\text{final}}$. In (a) the angular velocity increases over time, and in (b) the angular velocity decreases over time. In both cases, the angular acceleration vector $\vec{\alpha}$ points in the direction of the change in angular velocity $\Delta\vec{\omega}$.

We see from the definition that the SI unit is [rad/s^2], and also since Δt is a positive scalar, the average angular acceleration will have the same direction as $\Delta\vec{\omega}$. The **instantaneous angular acceleration** can be found by taking the limit as Δt goes to 0:

$$\vec{\alpha} = \lim_{\Delta t \to 0} \vec{\alpha}_{\text{avg}} = \lim_{\Delta t \to 0} \frac{\Delta\vec{\omega}}{\Delta t} = \frac{d\vec{\omega}}{dt}. \tag{1.29}$$

Of course, the units are the same as for the average angular acceleration and the direction is that of $d\vec{\omega}$.

In exactly the same way that the relationship between the directions of the linear velocity and linear acceleration vectors determine if an object speeds up or slows down linearly, if $\vec{\alpha}$ and $\vec{\omega}$ have the same direction, then the object is speeding up rotationally, but if they have opposite directions, the object is slowing down rotationally. These cases are shown schematically in Figure 1.11. In addition, just like we had discussed for the angular velocity, if the angular acceleration is constant, then the instantaneous angular acceleration is equal to the average angular acceleration:

$$\vec{\alpha}_{\text{avg}} = \vec{\alpha}$$

$$\Rightarrow \frac{\Delta\vec{v}}{\Delta t} = \frac{d\vec{v}}{dt}. \tag{1.30}$$

Lastly, we can find the relationship between the magnitude of an object's angular acceleration α and the magnitude of its tangential acceleration a_t based on what we have discussed so far. Since we know from the definition of linear acceleration that $a = \frac{dv}{dt}$, from Equation (1.16) ($v = \omega r$) we find:

$$a_t = \frac{dv}{dt} = \frac{d(\omega r)}{dt} = \frac{d\omega}{dt}r + \omega \underbrace{\frac{dr}{dt}}_{0} = \alpha r. \tag{1.31}$$

In this derivation, the second term in the sum goes to zero because if the object is rigid and the axis of rotation is fixed, r does not change, and thus $\frac{dr}{dt} = 0$. For the first term in the sum, we have used the definition $\frac{d\omega}{dt} = \alpha$. Equation (1.31) says that if an object is moving around a circle with a nonzero angular acceleration, then the object is speeding up or slowing down. Therefore, this acceleration can be expressed as either the linear acceleration ($\vec{a}_t$) tangent to the object's trajectory or the angular acceleration ($\vec{\alpha}$). By accounting for one, we have accounted for the other. However, whereas the angular acceleration is the same for all points on a rotating rigid body, the linear acceleration depends on the distance of the point from the axis of rotation. This is in complete analogy to what we learned about angular and tangential velocities.

1.3.3 Total Acceleration

If a rotating object speeds up or slows down, it has a tangential acceleration ($\vec{a}_t$). However, since it is moving in a circle, the object also has centripetal acceleration ($\vec{a}_c$). Both these accelerations have the same units and they can be added as vectors to obtain the total acceleration of the object, as shown in Figure 1.12.

The centripetal and tangential accelerations are perpendicular to each other, since the tangent at a point on a circle is perpendicular to the radius of the circle at that point. Then, the magnitude will be given by the Pythagorean theorem:

$$a_{\text{tot}} = \sqrt{a_t^2 + a_c^2}, \tag{1.32}$$

while the direction can be found via the equation:

$$\tan\theta = \frac{a_t}{a_c}. \tag{1.33}$$

Note that to these two accelerations $\vec{a}_t$ and $\vec{a}_c$, we cannot also add the angular acceleration $\vec{\alpha}$, since it has different units! We can only add the linear accelerations together. But more importantly, as we mentioned in Section 3.2, you do not need to worry – the tangential acceleration accounts for the angular acceleration. Therefore, we have already accounted for the angular acceleration in the sum of the two linear accelerations.

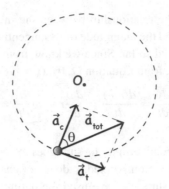

Figure 1.12 If an object is rotating, it has centripetal acceleration. If it is also speeding up or slowing down while executing rotational motion, it has a tangential acceleration. The two accelerations are perpendicular to each other and can be added together as vectors to find the total acceleration.

1.4 Rotational versus Linear Motion: Analogies

We now complete the analogy between translational and rotational motion, at least as far as the kinematic equations for constant velocity motion and constant acceleration motion are concerned. We see that the angular velocity and acceleration, since they were defined as the rotational counterparts of the linear velocity and acceleration, respectively, allow us to construct Table 1.1. For each of the cases shown in the table, the graphs of the angular physical quantities as functions of time will be the same as those of their translational counterparts. These graphs and the derivations of the equations for motion with constant angular velocity and constant angular acceleration will be the focus of one of the exercises at the end of the chapter.

Analogies like these shown in Table 1.1 are useful for developing an intuition for rotational motion. If at any point you are stuck solving a problem involving rotational quantities, think of the analogous problem involving linear quantities. At this stage you may find it easier to solve the linear motion problem, and you can then apply the same methodology to the rotational motion problem. Remember that for rotation about a fixed axis, all angular vector quantities are along a single axis of rotation, and thus the calculations involve only 1D vector addition, just as is the case in 1D translational motion.

Now, let's do an example that makes use of the analogies in Table 1.1.

Example 2: Spin rotational motion with constant angular acceleration. A wheel of radius $r = 2.0$ m spins with a constant angular acceleration of 3.5 rad/s^2 that causes the wheel to speed up. At time $t_0 = 0$ s, the angular velocity is 2.0 rad/s. (a) Through what angle does the wheel rotate between $t = 0$ s and

Table 1.1 *Physical quantities used to describe linear motion and their rotational motion counterparts.*

Linear Motion	Rotational Motion
Position $\vec{x}$	Angular position θ
Displacement $\Delta\vec{x}$	Angular displacement $\Delta\vec{\theta}$
Velocity $\vec{v} = \frac{d\vec{x}}{dt}$	Angular velocity $\vec{\omega} = \frac{d\vec{\theta}}{dt}$
Acceleration $\vec{a} = \frac{d\vec{v}}{dt}$	Angular acceleration $\vec{\alpha} = \frac{d\vec{\omega}}{dt}$
If $\vec{v}$ is constant: $\Delta\vec{x} = \vec{v}\Delta t$	If $\vec{\omega}$ is constant: $\Delta\vec{\theta} = \vec{\omega}\Delta t$
If $\vec{a}$ is constant:	If $\vec{\alpha}$ is constant:
$\vec{v} = \vec{v}_0 + \vec{a}\Delta t, \ \Delta\vec{x} = \vec{v}_0\Delta t + \frac{1}{2}\vec{a}\Delta t^2$	$\vec{\omega} = \vec{\omega}_0 + \vec{\alpha}\Delta t, \ \Delta\vec{\theta} = \vec{\omega}_0\Delta t + \frac{1}{2}\vec{\alpha}\Delta t^2$

$t = 2.0$ s? Give your answer in both radians and revolutions. (b) What is the angular speed of the wheel at $t = 2.0$ s? (c) What is the total acceleration of a point on the rim at time $t = 2.0$ s?

First, let's imagine the analogous problem in linear motion. It would go like this: "An object moves with constant linear acceleration of 3.5 m/s^2 so that it speeds up. At time $t = 0$ s, the velocity is 2.0 m/s. (a) What is the object's displacement between $t = 0$ s and $t = 2.0$ s? Give your answer in both meters and feet. (b) What is the speed of the object at $t = 2.0$ s?" Part (c) does not have an equivalent in linear motion.

How would you solve this new problem? It is not as difficult to develop a methodical solution in this case because you already have experience with linear motion problems. You can now use the same methodology to solve the equivalent rotational problem! We encourage you to try thinking about the analogous linear motion problems as you work through this book – it will help you build an intuition for rotational motion.

(a) Now, let's talk about our original spinning wheel and assume that the wheel rotates counterclockwise in the plane of the page, as shown in Figure 1.13. We define the $+z$ axis to be out of the plane of the page and perpendicular to it. Using the right-hand rule, we find that $\vec{\omega}$ is along the $+z$ direction and so is $\vec{\alpha}$, since the wheel is speeding up.

Based on the given information, we have rotational motion with constant angular acceleration $\vec{\alpha}$. The equations that describe this concept are:

$$\vec{\omega} = \vec{\omega}_0 + \vec{\alpha}\Delta t \tag{1.34}$$

and

$$\Delta\vec{\theta} = \vec{\omega}_0\Delta t + \frac{1}{2}\vec{\alpha}\Delta t^2. \tag{1.35}$$

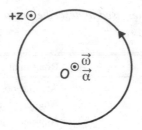

Figure 1.13 A wheel rotating with constant angular acceleration. The direction of rotation is counterclockwise when viewed from above, and so by the right-hand rule $\vec{\omega}$ points out of the page. $\vec{\alpha}$ points in the same direction as $\vec{\omega}$ since the wheel is speeding up.

We can plug the known parameters into equation (1.35) to get:

$$\Delta\vec{\theta} = \vec{\omega}_0 \Delta t + \frac{1}{2}\vec{\alpha}\Delta t^2$$

$$\Rightarrow \Delta\theta = 2.0 \text{ rad/s} \cdot 2.0 \text{ s} + \frac{1}{2} \cdot 3.5 \text{ rad/s}^2 \cdot (2.0 \text{ s})^2 = 11 \text{ rad}. \qquad (1.36)$$

Since 2π rad = 1 revolution, then the number of revolutions N will be

$$N = \frac{\Delta\theta}{2\pi} = \frac{11 \text{ rad}}{2\pi \text{ rad}} = 1.8 \text{ revolutions}. \qquad (1.37)$$

(b) To solve this part of the problem, we can plug the known values into equation (1.34):

$$\vec{\omega} = \vec{\omega}_0 + \vec{\alpha}\Delta t$$

$$\Rightarrow \omega = 2.0 \text{ rad/s} + 3.5 \text{ rad/s}^2 \cdot 2.0 \text{ s} = 9.0 \text{ rad/s}. \qquad (1.38)$$

(c) We can now easily find v at time $t = 2.0$ s. We have from Equation (1.16):

$$v = \omega r = 9.0 \text{ rad/s} \cdot 2.0 \text{ m} = 18 \text{ m/s}. \qquad (1.39)$$

Therefore, the centripetal acceleration is (Equation (1.25)):

$$a_c = \frac{v^2}{r} = \frac{(18 \text{ m/s})^2}{2.0 \text{ m}} = 162 \text{m/s}^2. \qquad (1.40)$$

The tangential acceleration is (Equation (1.31)):

$$a_t = \alpha r = 3.5 \text{ rad/s}^2 \cdot 2.0 \text{ m} = 7.0 \text{m/s}^2. \qquad (1.41)$$

As shown in Figure 1.12, these two are perpendicular to each other, so the magnitude of the total acceleration will be:

$$a_{\text{tot}} = \sqrt{a_t^2 + a_c^2} = \sqrt{(7.0\text{m/s}^2)^2 + (162\text{m/s}^2)^2} = 162.15\text{m/s}^2, \qquad (1.42)$$

while the direction can be found via the equation:

$$\tan \theta = \frac{a_t}{a_c} = \frac{7.0\text{m/s}^2}{162\text{m/s}^2} = 0.043 \Rightarrow \theta = 0.04 \text{ rad.} \qquad (1.43)$$

Exercises

(i) Think back to how you derived the equations of motion for 1D motion with constant velocity and 1D motion with constant acceleration. Now derive the equations of motion for rotational motion with constant angular velocity and rotational motion with constant angular acceleration, shown in Table 1.1. Also, construct graphs for the position, velocity, and acceleration (both angular and linear) as functions of time for all combinations of positive and negative velocity, and positive and negative acceleration. Compare the graphs for the linear and angular quantities as functions of time. Why should you not be surprised that the graphs look the same?

(ii) An object is moving counterclockwise around a circle of radius $r = 1.0$ m with constant speed. Within time $\Delta t = 10$ ms it traverses an angle of $120°$. Find (a) the object's angular and linear velocities for an arbitrary position of the object along the circle and (b) the period of the object's motion. (c) Draw a diagram to show the directions of the linear and angular velocities for an arbitrary position of the object along its trajectory.

(iii) While moving at a constant frequency of 0.5 Hz along the circumference of a circle of radius 0.5 m, an object passes through point A. At the same time, another object starts from rest from the same point A to move along the diameter AB. The second object moves with constant acceleration. (a) Find the second object's acceleration given the fact that the two objects meet at point B at the moment when the first object reached B for the third time (but the second object reached B for the first time). (b) What is each object's velocity at the moment they meet at point B?

(iv) Two cars are moving around the same circle of radius $R = 15$ m. Their linear, constant speeds are $v_1 = 3.0$ m/s and $v_2 = 2.0$ m/s. Assuming at time $t = 0$ the cars are at the same point along the circumference, find the first time that these two cars will meet again when (a) they move in the same direction around the circle and (b) they move in opposite directions around the circle.

(v) In Equation (1.31) we showed that $a_t = \alpha r$ by starting from Equation (1.16) and taking its time derivative. Instead, now use Equation (1.14) as the starting point. Take its time derivative to find the *total* acceleration.

(a) Show that the total acceleration consists of two terms:

  • The tangential acceleration $\vec{a}_t$ given by

$$\vec{a}_t = \vec{\alpha} \times \vec{r},\tag{1.44}$$

  • and the centripetal acceleration $\vec{a}_c$ given by the triple cross product

$$\vec{a}_c = \vec{\omega} \times (\vec{\omega} \times \vec{r}).\tag{1.45}$$

(b) Explain why the result of this triple cross product is consistent with the magnitude (Equation (1.25)) and the direction of the centripetal acceleration we discussed in Section 3.1. You might need to use the identity that $\vec{A} \times (\vec{B} \times \vec{C}) = \vec{B}(\vec{A} \cdot \vec{C}) - \vec{C}(\vec{A} \cdot \vec{B})$.

2

Torque and Static Equilibrium

As mentioned in Chapter 1, we know from linear motion and Newton's Second Law that to get an object to accelerate (i.e., change its linear velocity), we have to apply a force. Now we seek the equivalent of force in rotational motion. In other words, how can one get an object to change its angular velocity? The answer is that we must apply torque. Just like force was a ubiquitous concept in linear mechanics, torque is ubiquitous in rotational mechanics.

2.1 Torque ($\vec{\tau}$)

We begin this chapter with a definition and detailed discussion of torque, which we then use to study static equilibrium. In later chapters, we will use the concept of torque to introduce and discuss the dynamics of systems undergoing rotational motion.

2.1.1 Mass Attached to Massless, Rigid Rod

We begin with a thought experiment intended to provide some physical intuition about torque. Imagine we have a point mass m attached to the end of a massless but rigid rod (Figure 2.1). The rod is fixed at one end (i.e., point O) in such a way that the mass can only rotate about an axis that is perpendicular to the page and passes through point O. Because the rod is rigid, the distance between the mass m and point O is fixed and cannot change. Now, let's examine what happens when a force $\vec{F}$ is applied to the other end of the rod as shown in Figure 2.1.

Intuitively, we know that the resulting motion of the mass will depend on the direction at which $\vec{F}$ is applied. If the mass is initially at rest and a force is applied parallel to $\vec{r}$, the vector pointing from the axis of rotation to the point where the force is exerted on the mass, the mass will not accelerate in that direction since the system is fixed at point O (if the rod were not fixed at O,

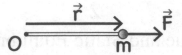

Figure 2.1 System of a mass m attached to the end of a massless rod. The rod is fixed at one end (i.e., point O) in such a way that the mass can only rotate about an axis that is perpendicular to the page and passes through point O. If the mass rotates, it will transcribe a circle with its center at O. A force $\vec{F}$ is exerted at the other end of the rod with $\vec{r}$ being the vector from point O to the point where the force is exerted. If the force $\vec{F}$ is applied parallel or anti-parallel to $\vec{r}$, then it will not cause a rotation about O. Furthermore, because the rod is fixed at point O, the force will not cause the system to accelerate to the right.

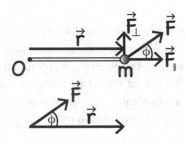

Figure 2.2 Decomposition of force $\vec{F}$ into components parallel ($\vec{F}_\parallel$) and perpendicular ($\vec{F}_\perp$) to $\vec{r}$. Only the perpendicular component $F_\perp = F \sin\phi$ will cause the mass to start rotating, where ϕ is the angle between $\vec{r}$ and $\vec{F}$.

then the force $\vec{F}$ would cause the system to accelerate to the right). Furthermore, this force will not cause a rotation. These outcomes will also be true if $\vec{F}$ is applied anti-parallel to $\vec{r}$. However, if the force is applied at any other angle, our intuition tells us that the mass will begin moving (i.e., accelerate), rotating about the fixed point.

To see why, it is necessary to examine the components of the force in relation to $\vec{r}$. As shown in Figure 2.2, $\vec{F}$ applied at some angle ϕ with respect to $\vec{r}$ can be decomposed into parallel ($\vec{F}_\parallel$) and perpendicular ($\vec{F}_\perp$) components. Since $\vec{F}_\parallel$ is parallel to $\vec{r}$, based on our previous discussion we can already conclude that this component of the force will not cause a rotation. However, since rotation does occur, we must then conclude that the component of $\vec{F}$ perpendicular to $\vec{r}$ is responsible for this rotation! Since $F_\perp = F \sin\phi$ we now know the magnitude of the component of the force that is responsible for the rotation. Of course, the greater $F \sin\phi$ is, the greater the acceleration of the mass (as will be discussed in Chapter 3).

Now let's think about the vector $\vec{r}$, the position vector that goes from the axis of rotation to the point where the force is exerted. The question is: Does

$|\vec{r}|$ matter? In other words, even if you have a perpendicular component of the force $\vec{F}_\perp$, will the mass-rod system rotate just as easily regardless of where the force is exerted along the rod?

Where the force is exerted does matter: The further away the same force is exerted from the axis of rotation, the easier it is to rotate the system. To gain some intuition, think about what happens when you try to open a door. If you exert a force at the axis of rotation (where the hinges are) where $|\vec{r}| = 0$, no matter how hard you push or pull, you will not get the door to open. However, the further away you exert the same force from the axis of rotation (i.e., as $|\vec{r}|$ increases), the easier it is to get the door to open. This is why the door handle is placed the farthest possible distance from the axis of rotation: The door will rotate more easily for the same amount of force if $|\vec{r}|$ is large (we will see what "easily" means in Chapter 3).

Therefore, we see that getting something to rotate depends on two things: The distance away from the axis of rotation that the force is exerted, $|\vec{r}|$, and the magnitude of the perpendicular component of the exerted force, $F \sin \phi$. It is important to emphasize that producing a rotation depends on all of these properties together; for example, we can have both nonzero $|\vec{r}|$ and $|\vec{F}|$ but still not have rotation if $\phi = 0$ or $\phi = \pi$.

2.1.2 Definition of Torque

In general, the torque $\vec{\tau}$ due to a force is the physical quantity that describes the ability a force has to rotate an object. Just like forces cause accelerations, torques cause angular accelerations, as we will see in Chapter 3.

In the previous section, we saw that the torque ($\vec{\tau}$) is related to the applied force ($\vec{F}$), the position vector pointing from the axis of rotation to the point of application of this force ($\vec{r}$), and the sine of the angle between these two vectors. We know that when a physical quantity is the result of two vectors which are related through the sine of the angle between them, the cross product is usually involved. This turns out to be the case here, as well. The torque due to a force is defined as:

$$\vec{\tau} = \vec{r} \times \vec{F}, \tag{2.1}$$

and has SI units of [Nm]. The magnitude is given by

$$|\vec{\tau}| = |\vec{r}||\vec{F}|\sin\phi, \tag{2.2}$$

which is consistent with what we saw intuitively in our case study.

Because torque is a cross product between the two vectors $\vec{r}$ and $\vec{F}$, it is also a vector whose direction is given by the right-hand rule. To find its direction,

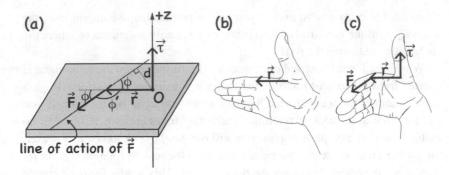

line of action of $\vec{F}$

Figure 2.3 (a) A rectangular slab can rotate about the fixed z axis that goes through point O and is perpendicular to the slab's plane. A force $\vec{F}$ is exerted at an angle ϕ with respect to $\vec{r}$. The moment arm, $r \sin \phi = d$, is the length of the line along the plane that starts from the axis of rotation and is perpendicular to the line of action of the force. Since $\phi' = \pi - \phi$, $\sin \phi' = \sin \phi$. (b) To find the direction of $\vec{\tau}$ using the right-hand rule, we first need to orient the four fingers of our right hand to point in the direction of the first vector in the cross product ($\vec{r}$). (c) Then, we bend our fingers to align them with the second vector in the cross product ($\vec{F}$) and our extended thumb points in the direction of the resulting vector ($\vec{\tau}$) (upward along the axis of rotation z).

we use the more general version of the right-hand rule discussed in Chapter 1. Let's review this rule in the context of torque by looking at Figure 2.3. Imagine that the rectangular slab shown in Figure 2.3a has the ability to rotate about the fixed z axis when a force $\vec{F}$ is exerted on it as shown. The force is located in the plane of the slab, which is perpendicular to the axis of rotation z. To get the direction of torque, we align the four fingers of our right hand with the first vector in the cross product ($\vec{r}$) (Figure 2.3b) in an orientation so that when we bend our fingers to align them with the second vector in the cross product ($\vec{F}$) (Figure 2.3c), our extended thumb points in the direction of the torque. In this example, the torque due to $\vec{F}$ points upward, and by convention is drawn along the axis of rotation (just like angular velocity and angular acceleration). As expected, we also see that $\vec{\tau}$ is perpendicular to both $\vec{F}$ and $\vec{r}$, as well as to the plane formed by $\vec{F}$ and $\vec{r}$ (i.e., the plane of the slab in this case). For the case studied in Figure 2.2, $\vec{\tau}$ will be pointing out of the page, toward the reader.

Let's look at a slightly different aspect of Figure 2.3. We can see from Figure 2.3a that

$$|\vec{r}| \sin \phi = d. \tag{2.3}$$

d corresponds to the perpendicular distance from the line of action of the force (i.e., the line that contains the vector $\vec{F}$) to the axis of rotation. d is called the

moment arm (or lever arm) and allows us to write the magnitude of the torque
(Equation (2.2)) as:

$$|\vec{\tau}| = |\vec{r}||\vec{F}|sin\phi = |\vec{F}|d. \tag{2.4}$$

Also, note that the angles ϕ and ϕ' in Figure 2.3a are supplementary; that
is, $\phi' = \pi - \phi$, and thus $\sin \phi' = \sin \phi$. As a result, one can compute the torque
using either the angle between the vectors $\vec{r}$ and $\vec{F}$, or its supplementary angle.

From this discussion and the one in the previous section (Section 1.1) we can
see that $\sin \phi$, when combined with the applied force $\vec{F}$, shows the component
of the force that causes rotation. However, when combined with the position
vector $\vec{r}$, $\sin \phi$ shows the moment arm. In other words, $\sin \phi$ can help us focus
our attention either on the applied force or on the position vector, depending on
the problem.

At this point you may notice that the SI unit for torque is the same as the
SI unit for work, [Nm]. Recall that work is defined as the integral of the dot
product of a force along an infinitesimal displacement:

$$W = \int_A^B \vec{F} \cdot d\vec{x}. \tag{2.5}$$

Torque, however, is the *cross* product between the position vector and the force,
and therefore it represents a different physical quantity. This is why we do not
use the unit of the Joule when expressing torque.

As simple as Equations (2.1) and (2.2) look, there are some important
observations that need to be made, and this leads us to the next section.

2.1.3 Notes on the Definition of Torque

Before we begin our observations, we need to remind ourselves that the focus
of this book is on an object's rotation about a *fixed* axis. This means that the
axis of rotation cannot rotate about another axis. Situations where the axis of
rotation rotates about a second axis (e.g., a spinning top that rotates about its
own axis while this axis rotates about an axis perpendicular to the tabletop)
will be briefly discussed in Chapter 8.

   (i) Since $\vec{r}$ is defined as the position vector from an axis to the point where
       the force is exerted, we always need to specify the axis about which we
       calculate the torque. For example, we need to say: "The torque of $\vec{F}_1$
       about the axis that goes through the object's center of mass (CM) and is
       perpendicular to the plane of the page."
   (ii) In Figure 2.3 we looked at the torque due to a force located in the plane
       of a rectangular slab. What happens if the force $\vec{F}$ is not located in this

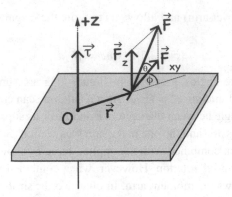

Figure 2.4 If a force has a component parallel to the axis of rotation (the z axis), that component ($\vec{F}_z$) cannot produce torque about the z axis, even if $\vec{F}_z$ is perpendicular to the vector $\vec{r}$. Torque about a particular axis can only be due to force components that are perpendicular to both the axis and to $\vec{r}$. Thus, in this case, only the component $F_{xy} \sin\phi = F\cos\theta \sin\phi$ produces torque about the z axis.

plane, but rather forms an angle θ with it, as shown in Figure 2.4? Then the torque about the axis of rotation (the z axis in this case) is given by the torque due to the component of this force which *is* perpendicular to the z axis, that is, the component of the force that lies in the xy plane of the slab, $\vec{F}_{xy}$. This makes sense because the component $\vec{F}_z$ of the force that is parallel to the axis of rotation cannot produce any rotation about this fixed axis. Of course, $\vec{F}_{xy}$ (with $F_{xy} = F\cos\theta$) can further be split into two components, one parallel and one perpendicular to $\vec{r}$. The perpendicular component of magnitude $F_{xy} \sin\phi = F\cos\theta \sin\phi$, where ϕ is the angle between $\vec{F}_{xy}$ and $\vec{r}$, is thus the only component that causes rotation about the fixed z axis. In summary, only the component of the force that is perpendicular to both the axis of rotation and $\vec{r}$ can produce a torque about that axis.

(iii) If we define the $+z$ axis to be upward along the axis of rotation (as is shown in Figure 2.4), then based on the right-hand rule, any force that tends to rotate the object counterclockwise (as viewed from above) will result in a torque in the +z direction. This results in the convention we will be using in this book: When a rotation is viewed from above the xy plane (i.e., such that $z > 0$), counterclockwise rotation is positive (just like the torque that causes it) while clockwise rotation is negative. For this reason, sometimes instead of showing the $+z$ axis in our figures, we will be showing the equivalent definition: That counterclockwise rotation is positive while clockwise rotation is negative. This convention is consistent with the right-hand rule. Of course, one could define the opposite convention.

(iv) In the same way we defined angular velocity and angular acceleration to point along the axis of rotation in Chapter 1, here by having all the torques be along the axis of rotation we are consistent with the ideas that the rotational motion can be fully described by vectors normal to the plane of rotation and that we are reducing the two-dimensional rotational motion to a one-dimensional problem. If we define the axis of rotation to be the z axis, some torques will be pointing in the $+z$ direction, while others in the $-z$ direction. This means that we can find the net torque in the same way we found the net force along an axis. That is, we consider the torques that point along the $+z$ axis positive (i.e., tending to cause counterclockwise (ccw) rotation) and those that point along the $-z$ axis negative (i.e., tending to cause clockwise (cw) rotation), and we are then able to add them algebraically. For example, for n torques:

$$\Sigma\vec{\tau} = \vec{\tau}_1 + \vec{\tau}_1 + ... + \vec{\tau}_n$$

$$= \underbrace{+\tau_1 + \tau_2 + \tau_3}_{\substack{\text{torques that would cause ccw} \\ \text{rotation (along } +z \text{ axis)}}} +... \underbrace{-\tau_{n-2} - \tau_{n-1} + \tau_n}_{\substack{\text{torques that would cause cw} \\ \text{rotation (along } -z \text{ axis)}}}$$

$$= +r_1F_1 \sin\phi_1 + r_2F_2 \sin\phi_2 + ... - r_{n-1}F_{n-1} \sin\phi_{n-1} - r_nF_n \sin\phi_n$$

$$= F_1d_1 + F_2d_2 + ... - F_{n-1}d_{n-1} - F_nd_n.$$

$$(2.6)$$

A word of caution about Equation (2.6): If we find the sign of the torques based on the direction of the rotation they would cause (clockwise or counterclockwise), then all the angles ϕ should be between 0 and π (so that $\sin\phi > 0$), since the sign of the torque has already been taken into account.

(v) If the point of application of force $\vec{F}$ changes location but the direction and line of action of the force are maintained (i.e., the vector remains constant but can be applied anywhere along its line of action), then its torque does not change. This is because the moment arm d will not change (i.e., $r \sin\phi$ will remain the same). As seen in Figure 2.5, $d = r_1 \sin\phi_1 = r_2 \sin\phi_2$, and thus the torque is the same.

(vi) The torque of a force $\vec{F}$ is zero ($\vec{\tau}_F = 0$) about a particular axis when:

- The line of action of $\vec{F}$ passes through that axis. This is because the moment arm d is zero (i.e., either $\phi = 0$ or $\phi = \pi$),
- The line of action of $\vec{F}$ is parallel to that axis. This was explained in note ii.

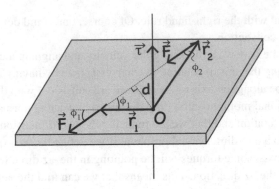

Figure 2.5 If the point of application of force $\vec{F}$ changes location but the direction and line of action of the force are maintained, then the moment arm ($d = r_1 \sin \phi_1 = r_2 \sin \phi_2$) does not change. Thus, the torque of the force remains the same.

(vii) When finding the torque of a force about an axis (let's assume the z axis), we can either work with the force $\vec{F}$ directly or with the components $\vec{F}_x$ and $\vec{F}_y$ of the force – both methods will yield the same result, that is $\vec{\tau}_F = \vec{\tau}_{Fx} + \vec{\tau}_{Fy}$. Let's see why this is true. Given that $\vec{F} = \vec{F}_x + \vec{F}_y$, we get that:

$$\vec{\tau} = \vec{r} \times \vec{F} = \vec{r} \times (\vec{F}_x + \vec{F}_y) = \vec{r} \times \vec{F}_x + \vec{r} \times \vec{F}_y = \vec{\tau}_{Fx} + \vec{\tau}_{Fy}. \quad (2.7)$$

Working with the components of the force is the preferred method in this book. This is because as soon as we decompose the force into components such that one is perpendicular to its respective position vector and the other parallel to it, then the angle ϕ in the definition of the torque $\tau = rF \sin \phi$ will be 0, $\frac{\pi}{2}$, or π, which are easy angles to work with. Decomposing the force might seem like an additional step. However, if we take this additional step, the physics can become clearer, particularly when the parallel and perpendicular directions lie along the axes of our coordinate system since we often look at the motion along each axis separately.

We have gone over a lot of theory. Time for some practice! Based on what we discussed in notes iii and iv, we can work with the torques algebraically when we account for their signs based on the direction of the rotation they tend to cause. As we do for translational motion, it is important to always show the coordinate system, draw the forces, decompose them, and write down the concepts and the equations that describe the concepts in each example.

Example 1: Net torque on a rod. Figure 2.6a shows a massless rod AC of length $l = 2.0$ m that can rotate in a horizontal plane about the z axis which is

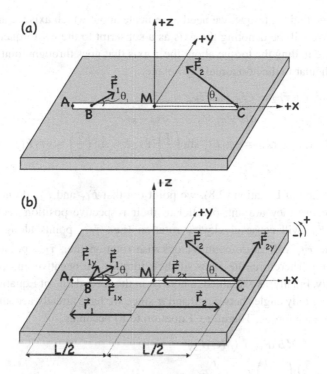

Figure 2.6 (a) The rod can rotate in the xy plane about the z axis which goes through its center M. The two forces $\vec{F}_1$ and $\vec{F}_2$ are exerted at different points along the rod and have different magnitudes and directions. (b) Forces $\vec{F}_1$ and $\vec{F}_2$ are broken into their x and y components. Position vectors are not drawn along the rod but are offset for clarity.

perpendicular to the plane and goes through the rod's midpoint M. Two forces are exerted along the horizontal plane, as shown. $\vec{F}_1$, which is exerted at point B, has a magnitude of 8.0 N, with $\theta_1 = 30°$ and $AB = l/8$. $\vec{F}_2$, which is exerted at point C, has a magnitude of $14\sqrt{2}$ N with $\theta_2 = 45°$. Find the net torque about the z axis.

For this problem, we will be using the concepts of torque and finding the net torque, Equations (2.2) and (2.6), respectively. It is important to remember that, *by definition*, ϕ is the angle between $\vec{r}$ and $\vec{F}$, **not** between $\vec{F}$ and either of the coordinate system axes. For the two forces $\vec{F}_1$ and $\vec{F}_2$ we will also work with their components, as shown in Figure 2.6b. In the same figure, we also indicate that counterclockwise rotation corresponds to positive torque while clockwise rotation corresponds to negative torque. As discussed in note iii, this notation is equivalent to saying that the $+z$ axis is upwards.

When we find the torque, we need to indicate about which axis we are evaluating it. We will be denoting the axis as a subscript to the net torque. In this case, we are finding the torque about the z axis that goes through point M and is perpendicular to the horizontal (xy) plane.

$$\Sigma \vec{\tau}_z = \vec{\tau}_{F1} + \vec{\tau}_{F2} = \underbrace{\vec{\tau}_{F_{1x}}}_{0\,(\phi\,=\,\pi)} + \vec{\tau}_{F_{1y}} + \underbrace{\vec{\tau}_{F_{2x}}}_{0\,(\phi\,=\,\pi)} + \vec{\tau}_{F_{2y}}$$

$$\Rightarrow \Sigma \tau_z = -\tau_{F_{1y}} + \tau_{F_{2y}} = -r_1 F_{1y} \sin\left(\frac{\pi}{2}\right) + r_2 F_{2y} \sin\left(\frac{\pi}{2}\right) = -r_1 F_{1y} + r_2 F_{2y}.$$
(2.8)

In the first line of Equation (2.8), we point out that $\vec{F}_{1x}$ and $\vec{F}_{2x}$ do not cause torque because they are anti-parallel to their respective position vectors. In addition, $\vec{F}_{1y}$ will cause clockwise rotation (i.e., $\vec{\tau}_{F_{1y}}$ points along the $-z$ axis) while $\vec{F}_{2y}$ will cause counterclockwise rotation (i.e., $\vec{\tau}_{F_{2y}}$ points along the $+z$ axis). Therefore, the signs of their torques are negative and positive, respectively. Finally, as explained in note iv, in the second line of Equation (2.8) we are using only angles between 0 and π since we have already accounted for the sign of each torque. Therefore, Equation (2.8) becomes:

$$\Sigma \tau_z = -(MB)F_{1y} + (MC)F_{2y}$$

$$= -\left(\frac{l}{2} - \frac{l}{8}\right) F_{1y} + \frac{l}{2} F_{2y}$$

$$= -\frac{3l}{8} F_1 \sin\theta_1 + \frac{l}{2} F_2 \sin\theta_2$$

$$= -\frac{3 \cdot (2.0\text{ m})}{8} \cdot (8.0\text{ N}) \sin\left(\frac{\pi}{6}\right) + \frac{2.0\text{ m}}{2} \cdot (14\sqrt{2}\text{ N}) \sin\left(\frac{\pi}{4}\right)$$

$$= -3\text{ Nm} + 14\text{ Nm} = +11\text{ Nm}.$$
(2.9)

The positive sign of our answer indicates that $\Sigma \vec{\tau}_z$ points along the $+z$ axis (i.e., the net force will produce counterclockwise rotation).

Example 2: Torque on a moving particle. A particle of mass m is moving along a trajectory in the xy plane and its coordinates at any time t are given by $(x, y) = (x_0 + \cos(\omega_1 t), y_0 + \sin(\omega_2 t))$, where x_0, y_0, ω_1, and ω_2 are constants. What is the torque on the particle about the origin at time $t = 0$?

Based on the x and y coordinates, the position $\vec{r}(t)$, velocity $\vec{v}(t)$, and acceleration $\vec{a}(t)$ of the particle are given by

$$\vec{r}(t) = (x_0 + \cos(\omega_1 t))\hat{i} + (y_0 + \sin(\omega_2 t))\hat{j},$$
(2.10)

$$\vec{v}(t) = \frac{d\vec{r}(t)}{dt} = -\omega_1 \sin(\omega_1 t)\hat{i} + \omega_2 \cos(\omega_2 t)\hat{j},$$
(2.11)

and

$$\vec{a}(t) = \frac{d\vec{v}(t)}{dt} = -\omega_1^2 \cos(\omega_1 t)\hat{i} - \omega_2^2 \sin(\omega_2 t)\hat{j}. \qquad (2.12)$$

At time $t = 0$, the above equations give that

$$\vec{r}(0) = (x_0 + 1)\hat{i} + y_0\hat{j}, \qquad (2.13)$$

$$\vec{v}(0) = \omega_2\hat{j}, \qquad (2.14)$$

$$\vec{a}(0) = -\omega_1^2\hat{i}. \qquad (2.15)$$

Since by Newton's Second Law $\Sigma\vec{F} = m\vec{a}$ and by definition $\vec{\tau} = \vec{r} \times \vec{F}$, we have

$$\begin{aligned}
\vec{\tau}(0) &= \vec{r}(0) \times \vec{F}(0) \\
&= \vec{r}(0) \times m\vec{a}(0) \\
&= m(\vec{r}(0) \times \vec{a}(0)) \\
&= m((x_0 + 1)\hat{i} + y_0\hat{j}) \times (-\omega_1^2\hat{i}) \\
&= m(x_0 + 1)(-\omega_1^2)\underbrace{\hat{i} \times \hat{i}}_{0} + my_0(-\omega_1^2)\underbrace{\hat{j} \times \hat{i}}_{-\hat{k}} \\
&= my_0(-\omega_1^2)(-\hat{k}) \\
&= my_0\omega_1^2\hat{k}. \qquad (2.16)
\end{aligned}$$

Our result makes sense. At all times the position vector is in the xy plane. The force vector, as seen from the acceleration, is also in the xy plane. The torque has to be perpendicular to each of these vectors and the plane formed by them. Thus, the torque is in the $\pm z$ direction, as we have shown mathematically

Example 3: Torque due to a continuous force. A horizontal wheel of radius $R = 1.0$ m has forces exerted at all points of its circumference. The forces are tangent to the circumference and act in the plane of the wheel, but the magnitudes of the forces vary as a function of the position, as shown in Figure 2.7. Starting from the point where the rim intersects the $+x$ axis and moving counterclockwise, the magnitude of the force is given as $F(\theta) = 2.0 - \sin\theta$ (N). Of course, θ can take values from 0 to 2π. Find the net torque about an axis that goes through the center of the wheel O and is perpendicular to its plane (the z axis).

For this problem, we will be using the concepts of torque and finding the net torque, Equations (2.2) and (2.6), respectively. We notice that for all forces, the magnitude of the position vector r is equal to the radius R and the angle between the position vector and the force is always $\phi = \pi/2$. Furthermore, all forces will produce counterclockwise rotation when viewed from above, so all the torques are positive. As a result, the net torque is given by:

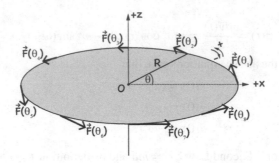

Figure 2.7 A wheel rotates due to tangent forces exerted at all points of its circumference. Though the angle between the force and the position vector is always $\pi/2$, the magnitude of the force varies along the circumference.

$$\Sigma \vec{\tau}_z = \vec{\tau}_1 + \vec{\tau}_2 + \vec{\tau}_3 + ...$$
$$\Rightarrow \Sigma \tau_z = +\tau_1 + \tau_2 + \tau_3 + ...$$
$$= +RF_1 \sin\left(\frac{\pi}{2}\right) + RF_2 \sin\left(\frac{\pi}{2}\right) + RF_3 \sin\left(\frac{\pi}{2}\right)...$$
$$= R(F_1 + F_2 + F_3 + ...) = R(\Sigma_i F_i). \tag{2.17}$$

It is clear at this point that we need to find the sum of all the forces' magnitudes. However, since the magnitude changes continuously with the position θ on the wheel, the sum becomes an integral where θ is the variable of integration. Therefore, the net torque is:

$$\Sigma \tau = R \int_0^{2\pi} (2 - \sin\theta)d\theta = R(2\theta + \cos\theta)|_0^{2\pi} = 1.0\text{m}(4\pi + 1 - 1)\text{ N} = 4\pi \text{ Nm} \tag{2.18}$$

in the +z direction.

2.2 Static Equilibrium of a Rigid Body

Now that we have defined the concept of torque and we have seen examples of how to calculate it, we will move to study the equilibrium of rigid bodies. As defined in Chapter 1, a rigid body is a solid object in which deformation under the application of a force is zero, or so small it can be neglected. For a rigid body, therefore, the distance between any two points of the object will not change over time.

Since we now have two types of motion to consider, translational and rotational, we have two types of equilibrium. In the case of translational motion, we recall that an object is in translational equilibrium when the net force on that object is equal to zero. We define rotational equilibrium in a similar way, namely, when the net torque on an object is zero. But why is this? In the case of translational motion, when the net force on an object is zero, the linear acceleration of the object is also zero, which means the object is either stationary or moving with constant linear velocity. We have said that torques are the equivalent of forces in the case of rotational motion. Then, when the net torque on an object is zero, the angular acceleration of the object is zero, which means the object is either not rotating or rotating with constant angular velocity (this will become more clear in the next chapter when we discuss Newton's Second Law for rotational motion). For an object to be in *static equilibrium*, therefore, it has to be in both translational and rotational equilibrium which means the following two conditions have to be satisfied:

(i) Translational Equilibrium:

$$\Sigma \vec{F} = 0 \Rightarrow \begin{cases} \Sigma \vec{F}_x = 0 \\ \Sigma \vec{F}_y = 0 \\ \Sigma \vec{F}_z = 0. \end{cases} \qquad (2.19)$$

(ii) Rotational Equilibrium:

$$\Sigma \vec{\tau} = 0. \qquad (2.20)$$

Equations (2.19) and (2.20) will be used every time we study an object in static equilibrium.

There is an important point that needs to be made here. If an object is stationary (or moving with constant angular velocity), its angular acceleration is zero. This means that no matter which axis we pick about which to calculate the torques on the object, the net torque $\Sigma \vec{\tau} = 0$. If the net torque were not equal to zero about even one axis, then there would be angular acceleration, and the object would not be stationary (or would not be moving with constant angular velocity). In many cases, we can pick an axis that will simplify the calculations, as we will see in our examples. This is why we did not include a subscript in Equation (2.20).

2.2.1 Drawing Forces

When drawing forces, it is important to consider the exact location of an object at which each force is exerted. Based on the definition of torque, the point of application of the force will determine the vector $\vec{r}$ and, therefore, the value of the torque. In this section, we focus on the direction and point of application of the following forces: The gravitational force, the force due to tension, the spring force, the reaction force from a supporting surface, the static friction force, the kinetic friction force, and the reaction force from a junction. We

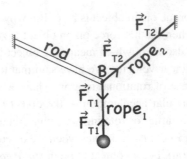

Figure 2.8 The system consists of two objects, the rod and the sphere, each supported by an inelastic rope. Rope 1 connects the sphere to the rod and rope 2 connects the rod to a wall. The sphere experiences the force due to tension $\vec{F}_{T1}$ while the rod experiences the forces due to tension of $\vec{F}'_{T1}$ and $\vec{F}_{T2}$. From Newton's Third law, $\vec{F}_{T1} = -\vec{F}'_{T1}$ and $\vec{F}_{T2} = -\vec{F}'_{T2}$.

expect you to know the magnitudes of forces such as the gravitational force, the spring force, and the frictional forces.[1]

- **The gravitational force $\vec{F}_g$:** The gravitational force $\vec{F}_g$ is exerted at the center of gravity of the object. The center of gravity coincides with the CM when the object is inside a uniform gravitational field, which is the case when close to the Earth's surface. Consequently, if an object is rotating about an axis that goes through its CM, $\vec{r} = 0$ and there is no torque due to $\vec{F}_g$. In fact, one can find the location of the CM of an extended object by computing the location about which the torque due to the gravitational force is zero. This is the focus of the first two exercises at the end of this chapter.

- **The force due to tension $\vec{F}_T$:** When an object is connected to a rope, then tension in the rope exerts a force $\vec{F}_T$ on the object along the rope. The point of application is where the object is connected to the rope. The magnitude of the force due to tension can vary depending on what other forces are exerted on the object. However, there is a maximum value that this force can take, above which the rope will break. In this book, we ignore the effects of elasticity. Figure 2.8 shows a sphere, a rod, and the two ropes that support them. Rope 1 exerts a tension force $\vec{F}_{T1}$ on the sphere. Both ropes 1 and 2 are connected to the rod at its end B. Thus, they exert forces $\vec{F}'_{T1}$ and $\vec{F}_{T2}$, respectively, on the end B. $\vec{F}'_{T2}$ is exerted by rope 2 on the wall.

- **The spring force $\vec{F}_s$:** The spring force $\vec{F}_s$ is exerted on the point of the object that is directly in contact with the spring. Recall that the direction of

---

[1] Wonderful descriptions of many of these forces are included in "A Student's Guide to Newton's Laws of Motion" by Sanjoy Mahajan, Cambridge University Press.

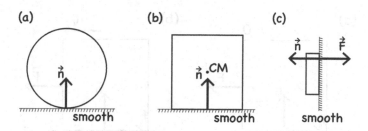

Figure 2.9 Any time an object is in contact with a surface, there is a reaction force $\vec{n}$ on the object at the point(s) of contact. (a) For a wheel on a surface, there is a single contact point. (b) For a box, there are multiple points of contact with the surface. However, we consider the gravitational force to be exerted at the center of mass (CM), essentially making the object dimensions irrelevant. As a result, the reaction force from the surface will oppose the gravitational force from the CM. (c) If a book is pushed against a wall via a force, then the reaction force is induced by that force. Thus, the reaction force will be exerted at the point of contact that is at the same height as the point at which the force $\vec{F}$ is applied.

the spring force is such that it always tries to return the spring to its natural length so that the spring is neither stretched nor compressed.

- **The reaction force from a supporting surface $\vec{n}$:** The reaction force $\vec{n}$ between an object and a surface is exerted on the point of the object that is in direct contact with the surface. For example, the reaction force on a wheel by a horizontal surface (Figure 2.9a) is exerted at the point of the wheel that is in contact with the surface. If there are more points between the two surfaces that are in contact (e.g., a box sliding on a table, as in Figure 2.9b), then the reaction force is exerted at the point of contact that is below the object's CM. This is because, as we just discussed, the gravitational force of an object acts at its CM. Therefore, if we consider all the mass to be concentrated at that single point, it is as if that point is in contact with the surface. Similarly, if we push a book against a wall (Figure 2.9c), the reaction force will be exerted at the point of contact with the wall that is at the same height as the point at which we exert the force. In summary, the reaction force "reacts" to oppose the force that wants to push the two surfaces together.

- **The static friction force $\vec{f}_s$:** In Figure 2.10a, we apply a force $\vec{F}$ on a box at rest on a rough surface. This external force attempts to move the box to the right. A force of static friction $\vec{f}_s$ will appear when the box and the rough surface attempt to slide against each other. The direction of this force is parallel to the two surfaces and against the intended direction of motion

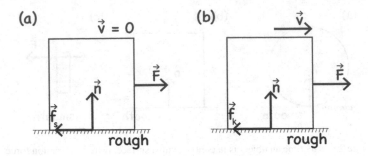

Figure 2.10 (a) If a stationary object has an external force $\vec{F}$ exerted on it, it will tend to move against the surface. The static friction force $\vec{f_s}$ exerted on the object will oppose the intended direction of motion. The reaction force $\vec{n}$ and the static friction $\vec{f_s}$ have the same point of application. The net force from the surface will be the vector sum of the two forces. (b) Once an object is moving against a rough surface, the kinetic friction $\vec{f_k}$ takes the place of the static friction $\vec{f_s}$. The net force from the surface will be the vector sum of the reaction force and the kinetic friction force.

of each surface. Thus, in Figure 2.10a the static friction on the box will be pointing to the left. We take its point of application to be the same as that of the reaction force $\vec{n}$. In summary, two forces are exerted on the box at the point of contact between the two surfaces: One is the reaction force $\vec{n}$ and the other is the static friction $\vec{f_s}$. The net force exerted by the surface on the box will be the vector sum of these two perpendicular forces.

- **The kinetic friction force $\vec{f_k}$:** We know from translational motion that when an object is moving against a rough surface, there is kinetic friction $\vec{f_k}$, which opposes the object's motion. Its point of application is the same as that of the reaction force $\vec{n}$. $\vec{f_k}$ is therefore now taking the place of $\vec{f_s}$ of the previous case, as shown in Figure 2.10b, and the net reaction force will be the vector sum of the reaction force and the kinetic friction force.

- **The reaction force from a junction $\vec{R}$:** When a junction connects an object to a firm surface, there is a reaction force $\vec{R}$ that is exerted from the firm surface on the object via the junction. This force has a magnitude and direction that varies, depending on what other forces are exerted on the object. However, the point of application of $\vec{R}$ on the object is the point of the object that is in contact with the junction. Figure 2.11 shows the forces exerted on a rod that remains horizontal with the help of a junction and a rope.

Now we are ready to do examples that illustrate these concepts. Because we are dealing with objects in equilibrium, the form of Equations (2.19) and (2.20) will require that we break forces into their components. If a force is

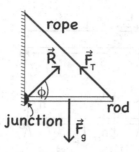

Figure 2.11 A rod remains horizontal and in equilibrium by having one end attached to a wall via a junction while the other end is also attached to the wall via a rope. The magnitude and direction of the reaction force $\vec{R}$ from the junction will depend on the gravitational force $\vec{F}_g$ and the force of tension $\vec{F}_T$.

unknown, then we will assume that all its components are positive (i.e., the components point in the positive direction of their respective axis) and we will draw them this way on our picture. The solution will tell us if one or more of the components point in the negative direction along their axis because our answer will have a negative sign. Furthermore, as mentioned, since we can pick any axis to be our axis of rotation for an object in equilibrium, we will pick our axis of rotation to go through a point at which one or more of the *unknown* forces are exerted. This will minimize the number of unknowns in Equation (2.20), which will simplify the algebra, as we will show in our next example.

Example 4: A system of objects in static equilibrium. The uniform, thin rod shown in Figure 2.12a has length $L = 8.0$ m and mass $M = 100$ kg. A small ball of $m = 20$ kg is placed $L_1 = 2.0$ m away from one end. Given that the gravitational acceleration is $g = 10 \, \text{m/s}^2$, find the reaction forces on the rod at points A and B where the rod is supported.

As always, we first draw all the forces. Because we have two objects of interest, the rod and the ball, we draw all forces on these two objects and break them into their components, if applicable. In this case, as seen in Figure 2.12b, all forces are along the y axis, so we do not need to decompose them. Of course, we also show the coordinate system and the positive and negative directions for rotation, as discussed previously.

The force $\vec{n}_{rb}$ is the force exerted on the ball by the rod while the force $\vec{n}_{br}$ is the force exerted on the rod by the ball. These two forces form an action–reaction pair and according to Newton's Third law $\vec{n}_{rb} = -\vec{n}_{br}$.

Since we have two objects, we will have to look at each object separately. For the ball, we can assume that the axis of rotation goes through its center O

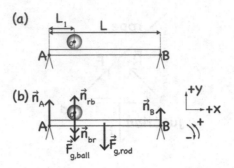

Figure 2.12 (a) A rod remains horizontal and in equilibrium via two supports, one on each end. A ball of mass m is located a distance L_1 away from end A. (b) All the forces exerted on the rod and the ball are along the y axis.

and is perpendicular to the plane of the page. We have two forces exerted on the ball, both along the y axis. So, our equations for equilibrium (Equations (2.19) and (2.20)) become, respectively:

$$\Sigma \vec{F}_y = 0$$

$$\Rightarrow \vec{n}_{rb} + \vec{F}_{g,\text{ball}} = 0$$

$$\Rightarrow n_{rb} - F_{g,\text{ball}} = 0$$

$$\Rightarrow n_{rb} = mg = 20 \text{ kg} \cdot 10 \text{ m/s}^2 = 2.0 \times 10^2 \text{ N} \tag{2.21}$$

and

$$\Sigma \vec{\tau}_O = 0 \Rightarrow \vec{\tau}_{n_{rb}} + \vec{\tau}_{F_{g,\text{ball}}} = 0. \tag{2.22}$$

However, each torque individually is equal to zero. The first because $\vec{F}_g$ is exerted at the axis of rotation, so its position vector is zero. The second because $\vec{n}_{rb}$, which is exerted at the point of contact between the ball and the rod, is antiparallel to its downward position vector. Therefore, Equation (2.22) does not give us any useful results.

Based on our result above we see that $\vec{n}_{rb} = -\vec{F}_{g,ball}$ and, from Newton's Third law, we also know that $\vec{n}_{rb} = -\vec{n}_{br}$. Therefore,

$$\vec{n}_{br} = \vec{F}_{g,\text{ball}}. \tag{2.23}$$

This result, though simple, is very subtle. It is common practice (even in many textbooks) to not write the equilibrium equations for the ball when solving problems like this. Most of the time, the equilibrium equations are only applied to the rod, which requires the assumption that the reaction force exerted by the ball on the rod is simply the gravitational force of the ball pushing down on the rod. This assumption cannot be made without justifying it, as we have done here, because the gravitational force of the ball is exerted on the ball, not on the rod. If one studies the rod, then the forces that need to be included are those on the rod, not on the ball. For equilibrium, this shortcut of assuming that $\vec{F}_{g,\text{ball}}$ is exerted directly on the rod (and jumping straight to Equation (2.25)) will work and we can take it, though it is not recommended because it cannot be applied to nonequilibrium situations. For this reason, we always look at each object separately when studying a system of objects and apply fundamental equations (like Newton's Third law in this case) to find out relationships between them.

We are now ready to study the rod. Again we start from the equilibrium Equations (2.19) and (2.20), keeping in mind that there are no force components along the x and z axes.

$$\Sigma \vec{F}_y = 0 \Rightarrow n_A - n_{br} - F_{g,\text{rod}} + n_B = 0. \tag{2.24}$$

But, as we have just discussed, $n_{br} = F_{g,\text{ball}} = 2.0 \times 10^2$ N. Therefore, the above equation becomes:

$$\Sigma \vec{F}_y = 0 \Rightarrow n_A - F_{g,\text{ball}} - F_{g,\text{rod}} + n_B = 0. \tag{2.25}$$

Rearranging and plugging the known values into Equation (2.25) gives:

$$n_A + n_B = F_{g,\text{ball}} + F_{g,\text{rod}}$$
$$= mg + Mg$$
$$= 20 \text{ kg} \cdot 10 \text{ m/s}^2 + 100 \text{ kg} \cdot 10 \text{ m/s}^2 = 1.2 \times 10^3 \text{ N}. \tag{2.26}$$

Now let's look at the rotational equilibrium for the rod starting from Equation (2.20). Since the rod is stationary, one can apply the condition for rotational equilibrium to any axis. We will pick an axis that goes through point A and is perpendicular to the plane of the page (we will explain why this is a good choice shortly). Given this axis, we have:

$$\Sigma \vec{\tau}_A = 0 \Rightarrow \vec{\tau}_{n_A} + \vec{\tau}_{n_{br}} + \vec{\tau}_{F_{g,\text{rod}}} + \vec{\tau}_{n_B} = 0. \tag{2.27}$$

Since we picked the axis of rotation to be through point A, $\vec{\tau}_{n_A} = 0$ because $\vec{r}_{n_A} = 0$ ($\vec{n}_A$ is exerted at point A). Since $\vec{n}_A$ is one of the unknowns, we would not know the value of $\vec{\tau}_{n_A}$ if we had not picked the axis of rotation to go through A. Therefore, by picking our axis of rotation in this way, we have eliminated one of the unknowns in Equation (2.27). The other unknown is $\vec{\tau}_{n_B}$, but we can now easily solve for it and, consequently, for n_B since this is the only unknown in our equation. Using our sign convention for the torques, we have:

$$\Sigma \tau_A = -\tau_{n_{br}} - \tau_{\text{Fg,rod}} + \tau_{n_B} = 0$$

$$\Rightarrow -L_1 n_{br} \sin\left(\frac{\pi}{2}\right) - \frac{L}{2} F_{\text{g,rod}} \sin\left(\frac{\pi}{2}\right) + L n_B \sin\left(\frac{\pi}{2}\right) = 0$$

$$\Rightarrow L_1 n_{br} \sin\left(\frac{\pi}{2}\right) + \frac{L}{2} F_{\text{g,rod}} \sin\left(\frac{\pi}{2}\right) = L n_B \sin\left(\frac{\pi}{2}\right)$$

$$\Rightarrow 2.0 \text{ m} \cdot 2.0 \times 10^2 \text{ N} + \frac{8 \text{ m}}{2} \cdot 1.0 \times 10^3 \text{ N} = (8.0 \text{ m}) n_B$$

$$\Rightarrow n_B = 5.5 \times 10^2 \text{ N}. \tag{2.28}$$

Since we had found that $n_A + n_B = 1.2 \times 10^3$ N, we see that $n_A = 6.5 \times 10^2$ N.

Because $\vec{n}_B$ is also unknown, one could have just as cleverly picked the axis of rotation to go through point B. In that case, $\vec{\tau}_{n_B} = 0$ because $\vec{r}_{n_B} = 0$ and we could solve Equation (2.27) for n_A instead. We should point out that we could also pick an axis of rotation that goes through the rod's midpoint or any other point. For example, had we picked the midpoint, where $\vec{F}_{\text{g,rod}}$ is exerted, $\vec{\tau}_{\text{Fg,rod}} = 0$ and there would be two unknowns $\vec{\tau}_{n_A}$ and $\vec{\tau}_{n_B}$ in Equation (2.27). We would still be able to solve for n_A and n_B in this case, by combining Equations (2.26) and (2.27), but the algebra would be a little more cumbersome. In many cases, the choice of the axis about which to compute the net torque will not prevent you from finding the solution – it will only help you by simplifying the algebra if you make a "smart" choice by picking the axis to go through a point where one of the unknown forces is exerted.

Example 5: Hanging by a rope. The uniform rod shown in Figure 2.13a has mass $M = 30$ kg and is kept horizontal with the help of a massless, inelastic rope and a junction. Given the angles shown in Figure 2.13a and that the gravitational acceleration is $g = 10$ m/s^2, find the reaction force from the junction and the force due to tension supplied by the rope.

Here we are interested in studying the rod, so let's first draw all the forces acting on the rod, as shown in Figure 2.13b. Because the rod is uniform, $\vec{F}_g$ is exerted at its midpoint. We note that both $\vec{F}_T$ and $\vec{R}$ are unknown forces. However, the direction of $\vec{F}_T$ is known (i.e., along the rope, toward the support point on the wall) while the direction of $\vec{R}$ is not known. For this reason, we will assume that the x and y components of $\vec{R}$ are positive and that $\vec{R}$ forms an angle θ with the $+x$ axis. Of course, we also draw our coordinate system.

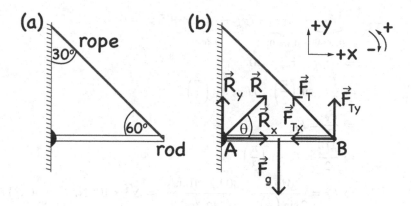

Figure 2.13 (a) A rod remains horizontal and in equilibrium while attached to the wall via a junction and a rope. The rod, the wall, and the rope form a $30° - 60° - 90°$ triangle. (b) The forces exerted on the rod are $\vec{F}_g$, $\vec{F}_T$ and $\vec{R}$. The directions of $\vec{F}_g$ and $\vec{F}_T$ are known, but $\vec{R}$ has an unknown direction. We assume that its x and y components are positive.

Starting from Equations (2.19) and (2.20), we obtain Equations (2.29), (2.30), and (2.31):

$$\Sigma \vec{F}_x = 0$$
$$\Rightarrow R_x - F_{Tx} = 0$$
$$\Rightarrow R_x = F_T \cos\left(\frac{\pi}{3}\right) = \frac{F_T}{2}, \tag{2.29}$$

and

$$\Sigma \vec{F}_y = 0$$
$$\Rightarrow R_y - F_g + F_{Ty} = 0$$
$$\Rightarrow F_g = R_y + F_T \sin\left(\frac{\pi}{3}\right) = R_y + \frac{\sqrt{3}}{2} F_T. \tag{2.30}$$

There are no force components along the z axis, so $\Sigma \vec{F}_z = 0$ is automatically satisfied.

For the rotational equilibrium, we pick our axis of rotation to be perpendicular to the plane of the page and go through point A. We pick point A since the unknown force $\vec{R}$ is exerted at that point. Based on our discussion in Example 4, another convenient choice would be an axis of rotation that goes through point B, since the force of tension is also unknown. We assume that l is the length of the rod, and thus for the net torque about an axis through point A we have:

$$\Sigma \vec{\tau}_A = \underbrace{\vec{\tau}_{R_x}}_{0,\,(r=0)} + \underbrace{\vec{\tau}_{R_y}}_{0,\,(r=0)} + \vec{\tau}_{Fg} + \underbrace{\vec{\tau}_{F_{Tx}}}_{0,\,(\phi\,=\,\pi)} + \vec{\tau}_{F_{Ty}} = 0$$

$$\Rightarrow -\tau_{Fg} + \tau_{F_{Ty}} = 0$$

$$\Rightarrow \frac{l}{2}F_g \sin\left(\frac{\pi}{2}\right) = lF_{Ty} \sin\left(\frac{\pi}{2}\right)$$

$$\Rightarrow \frac{F_g}{2} = F_{Ty}$$

$$\Rightarrow \frac{Mg}{2} = F_T \sin\left(\frac{\pi}{3}\right)$$

$$\Rightarrow F_T = \frac{Mg}{2\sin\left(\frac{\pi}{3}\right)} = \frac{30\text{ kg} \cdot 10\text{m/s}^2}{2\frac{\sqrt{3}}{2}} = \sqrt{3} \times 10^2 \text{ N}. \qquad (2.31)$$

Since now we know the magnitude of the force due to tension, we can find the values of R_x and R_y from Equations (2.29) and (2.30), respectively:

$$R_x = \frac{F_T}{2} = \frac{\sqrt{3} \times 10^2 \text{ N}}{2} = 50\sqrt{3}\text{ N}, \qquad (2.32)$$

$$R_y + \frac{\sqrt{3}}{2}F_T = Mg$$

$$\Rightarrow R_y = Mg - \frac{\sqrt{3}}{2}F_T = 30\text{kg} \cdot 10\text{ m/s}^2 - \frac{\sqrt{3}}{2} \cdot \sqrt{3} \times 10^2 \text{ N} = 1.5 \times 10^2 \text{ N}. \tag{2.33}$$

Both components are found to be positive, which means our initial assumption about $\vec{R}$ having its components point along the positive x and y axes is correct. If one or both of the components were negative, then the negative component(s) would point along the negative direction of that axis.

For the magnitude and direction of $\vec{R}$, we have:

$$R = \sqrt{R_x^2 + R_y^2} = \sqrt{(50\sqrt{3}\text{ N})^2 + (1.5 \times 10^2 \text{ N})^2} = \sqrt{3} \times 10^2 \text{ N}, \quad (2.34)$$

$$\tan\theta = \frac{R_y}{R_x} = \frac{1.5 \times 10^2 \text{ N}}{50\sqrt{3}\text{ N}} \Rightarrow \theta = \frac{\pi}{3} \text{ rad}. \qquad (2.35)$$

Of course, there is a second solution to Equation (2.35): $\theta = \frac{4\pi}{3}$ rad. But, since our results show that both $R_x, R_y > 0$, the angle θ belongs in the first quadrant. So $\theta = \frac{4\pi}{3}$ rad is not a physically acceptable solution.

2.3 Some Cool Theorems

There are some interesting theorems that can prove useful when working with problems on rotational equilibrium, but these theorems are not needed to solve problems. We mentioned in the preface to this book that we will not be relying

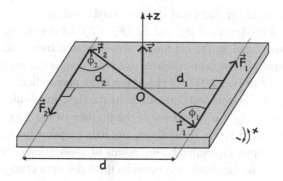

Figure 2.14 When two forces have equal magnitudes, opposite directions, and lines of action that do not overlap, as is the case with $\vec{F}_1$ and $\vec{F}_2$, they are known as a force couple. The net torque due to this couple has a magnitude equal to the magnitude of one force times the distance between the two forces.

on shortcuts to solve problems, and indeed, we have not used any in the examples we have solved so far – we have always started from fundamental equations. We present these theorems at the end of this chapter for the sake of completeness of this unit on torque and equilibrium, and because it is beneficial to know that they exist in the literature.

2.3.1 Theorem 1: Torque due to a Force Couple

A force couple is defined as a system of two forces that have

(i) Equal magnitudes,
(ii) Opposite directions,
(iii) Lines of action that do not overlap.

In Figure 2.14, $\vec{F}_1$ and $\vec{F}_2$ are a force couple. A force couple is not to be confused with an action-reaction pair of forces described by Newton's Third Law. An action-reaction pair corresponds to the forces that two objects exert on each other: For example, object 1 acts via a force on object 2, and object 2 reacts by exerting a force of equal magnitude and opposite direction on object 1. On the other hand, $\vec{F}_1$ and $\vec{F}_2$ in Figure 2.14 are both exerted by an external agent on the object of interest.

Let's find the net torque due to this force couple about the z axis. Both forces produce positive torques based on the right-hand rule (i.e., each will cause counterclockwise rotation):

$$\Sigma \vec{\tau}_z = \vec{\tau}_1 + \vec{\tau}_2$$

$$\Rightarrow \Sigma \tau_z = +r_1 F_1 \sin \phi_1 + r_2 F_2 \sin \phi_2$$

$$= +F_1 d_1 + F_1 d_2 = +F_1(d_1 + d_2) = +F_1 d, \qquad (2.36)$$

where we have used the facts that $d_1 = r_1 \sin \phi_1$ and $d_2 = r_1 \sin \phi_2$ are the moment arms, with $d = d_1 + d_2$, and that $F_1 = F_2$. Of course, had the forces caused clockwise rotation, the net torque would be negative in our coordinate system. Consequently, the net torque due to a force couple has magnitude equal to $\Sigma \tau_z = F_1 d = F_2 d$ (since $F_1 = F_2$) and direction perpendicular to the plane formed by the force couple and determined by the right-hand rule.

Note 1: Since $d_1 + d_2 = d$ regardless of the position of the z axis, the torque due to a force couple is the same about any axis perpendicular to the plane formed by the couple. Equivalently, the points of application of the forces do not matter, as long as the distance d between them does not change.

Note 2: By the definition of a force couple:

$$\Sigma \vec{F} = \vec{F}_1 + \vec{F}_2 = 0. \tag{2.37}$$

Therefore, a force couple only affects the object's rotational motion and does not affect the object's translational motion.

2.3.2 Theorem 2: Forces with Parallel Lines of Action

If an object is in equilibrium with n forces acting on it, and the lines of action of $n-1$ forces are parallel to each other, then the line of action of the nth force is also parallel to the other forces.

To see why this is the case, let's assume that all the $n-1$ parallel forces are along the y axis. Then the nth force also has to be along the y axis for the object to be in equilibrium. If the nth force is not along the y axis, then there will be a component of that force along a different axis (say the x axis); since there are no other forces along the x axis, then the object will not be in equilibrium along the x axis.

An example of how this theorem is applied is shown in the scenario of Figure 2.15, which is similar to one of the exercises at the end of this chapter. Figure 2.15 shows a rod that is attached to a wall via a junction and remains horizontal with the help of a spring. The forces exerted on the rod are the reaction force $\vec{R}$ from the junction, the gravitational force $\vec{F}_g$, and the spring force $\vec{F}_s$. Both $\vec{F}_g$ and $\vec{F}_s$ are along the y axis. Therefore, for the rod to remain in equilibrium, the reaction force from the junction $\vec{R}$ also has to be along the y axis. Whether it points along the positive or negative y direction is determined by the relative strength of $\vec{F}_g$ and $\vec{F}_s$. For example, if $F_g > F_s$ then $\vec{R}$ needs to be pointing upwards so that together with $\vec{F}_s$ they can cancel out $\vec{F}_g$.

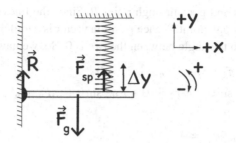

Figure 2.15 The spring force and the gravitational force on the rod are along the y axis. This means that, for the rod to be in equilibrium, the reaction force $\vec{R}$ from the junction also has to be along the y axis and it cannot have any x components. If $R_x \neq 0$, then the rod would not be in equilibrium along the x axis since there are no other forces along this axis to cancel out $\vec{R}_x$.

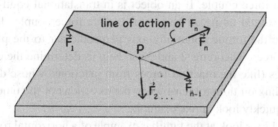

Figure 2.16 The slab is in equilibrium with the help of n forces. If the lines of action of forces $\vec{F}_1$, $\vec{F}_2$, ...$\vec{F}_{n-1}$ all intersect at point P, then the line of action of the nth force also has to intersect the lines of action of the other forces at the same point P.

2.3.3 Theorem 3: Forces Whose Lines of Action Intersect

If an object is in equilibrium with n forces acting on it, and the lines of action of $n-1$ forces intersect at the same point P, then the line of action of the nth force will also intersect the lines of action of the other forces at the same point P.

The proof of this theorem goes as follows: Let's look at the slab shown in Figure 2.16. This slab is in equilibrium under the influence of n forces, all of which lie on the same plane (i.e., the plane of the slab). The lines of action of $\vec{F}_1, \vec{F}_2, ...\vec{F}_{n-1}$ all intersect at point P.

Since the object is in equilibrium, we can pick any axis of rotation and apply Equation (2.20). For simplicity, let's pick an axis that is perpendicular to the plane of the forces and goes through point P. Since the lines of action of the $n-1$ forces pass through point P, each position vector is parallel to its respective force vector, and so the angle between the two is 0. So, we have:

$$\Sigma \vec{\tau}_P = 0 \Rightarrow \vec{\tau}_1 + \vec{\tau}_2 + \vec{\tau}_3 + ...\vec{\tau}_{n-1} + \vec{\tau}_n = 0$$

$$\Rightarrow F_1 r_1 \sin 0 + F_2 r_2 \sin 0 + F_3 r_3 \sin 0 + ...F_{n-1} r_{n-1} \sin 0 + \tau_n = 0$$

$$\Rightarrow \tau_n = 0$$

$$\Rightarrow F_n r_n \sin \phi = 0$$

$$\Rightarrow \sin \phi = 0. \tag{2.38}$$

Therefore, the line of action of $\vec{F}_n$ passes through point P also. Of course the derivation would be similar if some of the position vectors were antiparallel to their respective force vectors.

But, why are these theorems cool? Theorem 1 allows us to quickly find the torque due to a force couple. If an object is in translational equilibrium, then all the forces exerted on it can be grouped into a force couple. Then we can easily calculate the torque about any axis perpendicular to the plane formed by these two forces. Theorems 2 and 3 can help us determine the directions of unknown forces (like the reaction forces from junctions, whose direction can change depending on the problem) when the directions of the other forces are known. Let's quickly look at one example.

In Figure 2.17 we look at the familiar example of a horizontal rod in equilibrium attached to a wall via a junction and a rope. When we studied this system before (Example 5), we found the magnitude and direction of the reaction force $\vec{R}$ from the junction using the equations for static equilibrium. However, Theorem 3 can easily show us the direction of the force. Since the tension force and

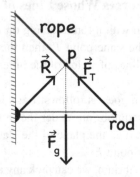

Figure 2.17 When the rod is in equilibrium, the direction of the reaction force $\vec{R}$ from the junction can be found by applying Theorem 3: If the force due to tension and gravitational force have lines of action that intersect at point P, then the line of action of $\vec{R}$ must go through point P also.

the gravitational force have lines of action that intersect at point P, the line of action of $\vec{R}$ also has to go through point P (but it will be up to us to determine if it is pointing toward or away from P).

Lastly, it is important to remember that Theorems 2 and 3 are applicable only if the object in question is in equilibrium.

Exercises

(i) The location of the CM of an extended object can be found by computing the location of the axis about which the torque due to the gravitational force is zero. (a) To show this, first consider a one-dimensional rod of length L and mass M placed along the x axis on this page. Let's pick the origin of the coordinate system to be on the left end of the rod and split the rod into infinitesimal masses dm each with its own position x with respect to the origin and its own gravitational force $\vec{F}_g$ along the y axis on this page. Each of these gravitational forces will result in a torque about an axis going through the origin of your coordinate system and perpendicular to the plane of the page. If we were to replace the extended object with a point mass M, where should this mass be located with respect to the origin so that we will get the same torque about the axis through the origin due to the gravitational force on this mass $\vec{F}_{g_M}$? This location is defined as the CM. (b) Now that we found the location of the CM with respect to the origin, show that the net torque due to the gravitational forces on the infinitesimal masses dm about an axis that goes through the center of mass of the rod is zero.

(ii) Imagine we have a horizontal rod supported either directly below or directly above its CM as shown in Figure 2.18a and 2.18b, respectively. Now, let's tilt the rod slightly and then release it. In the case where the support is below the rod, the rod will continue to tilt and will eventually

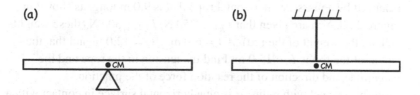

Figure 2.18 A rod is supported from directly below or directly above its center of mass (CM). (a) A small tilt will cause the rod to fall. (b) The rod will return to its horizontal position after a tilt.

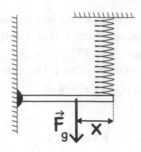

Figure 2.19 A rod is kept horizontal via a junction and a spring. The gravitational force $\vec{F}_g$ is exerted at a distance $x = 40$ cm from the end of the rod where the spring is attached.

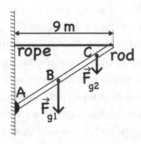

Figure 2.20 A rod is kept balanced via a junction and a massless, horizontal rope. The gravitational force on the rod is given as two separate vectors, $\vec{F}_{g1}$ and $\vec{F}_{g2}$.

fall. However, if the rod is supported from directly above its CM, it will always return to its original, horizontal position. Explain why this is true.

(iii) One end of a rod of length $L = 1.0$ m is attached to a wall via a junction that is free to rotate. The rod is kept horizontally balanced via a massless spring attached to its other end, as shown in Figure 2.19. You are given that $F_g = 100$ N and the CM is $x = 40$ cm away from its right end. Find the force exerted by the spring and the reaction force of the junction.

(iv) A rod is attached to a wall via a junction that is free to rotate and is kept balanced by a massless, horizontal rope that is 9.0 m long, as shown in Figure 2.20. You are given that $F_{g1} = 750$ N, $F_{g2} = 300$ N (these weights include the weight of the rod), $AB = 6.0$ m, $AC = 12.0$ m and that the length of the rod is $L = 15.0$ m. Find the tension in the rope and the magnitude and direction of the reaction force of the junction.

(v) A uniform wheel with radius a is on a horizontal surface in contact with a step of height h as shown in Figure 2.21. The gravitational force $\vec{F}_g$ is exerted at the wheel's center O. A force $\vec{F}$ is exerted at point Z, which is

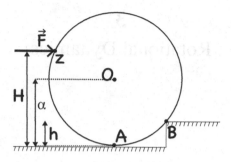

Figure 2.21 We exert a horizontal force $\vec{F}$ on a uniform wheel of radius α to roll it up a step of height h.

at a height H above the horizontal surface, and causes the wheel to roll up the step. (a) Draw all the forces exerted on the wheel before it rolls up the step. What happens to the ground's reaction force $\vec{n}_A$ on the wheel at point A when the wheel starts rolling up the step and loses contact with the ground? (b) Find the minimum magnitude of $\vec{F}$ needed for the wheel to roll up the step. (c) Find the direction of the step's reaction force $\vec{R}$ on the wheel at point B in two different ways: (i) using Newton's Laws and (ii) using geometry and Theorem 3.

3

Rotational Dynamics

In Chapter 1, we studied kinematics in the context of rotational motion and learned that there are many parallels to linear (or translational) kinematics. In this chapter, we will develop techniques to study the dynamics of rotating systems by building upon what we already know about torque from Chapter 2 and forces in translational motion. When studying rotational kinematics in Chapter 1, we were given the angular velocity or the angular acceleration but were not concerned with their physical cause. In this chapter, we will discuss how to produce an angular acceleration and thus change the angular velocity.

3.1 Newton's Second Law for Rotational Motion

Let's remind ourselves of Newton's Second Law for translation motion:

$$\Sigma \vec{F} = m\vec{a} \Rightarrow \vec{a} = \frac{\Sigma \vec{F}}{m}. \tag{3.1}$$

As mentioned previously in Chapter 2, Equation (3.1) says that a net force $\Sigma \vec{F}$ causes a mass m to accelerate with an acceleration of $\vec{a}$. The mass m expresses how difficult it is to get an object to change its velocity (i.e., accelerate). Given the same $|\Sigma \vec{F}|$, a greater value of m results in a smaller value of $|\vec{a}|$. Also, given that m is a positive scalar quantity, the direction of the acceleration $\vec{a}$ will be that of the net force $\Sigma \vec{F}$.

We will now study how to apply Newton's Second Law to objects that execute rotational motion. We start by referring back to the case study at the beginning of Chapter 2: A point mass m attached to the end of a massless but rigid rod (shown again in Figure 3.1). The rod is fixed at one end (i.e., point O) in such a way that the mass can only rotate about point O in the plane of the page, with the axis of rotation perpendicular to the page and passing through O. We apply a force $\vec{F}$ to m which forms an angle ϕ with the position vector $\vec{r}$ (the vector pointing from the axis of rotation to the point of application of $\vec{F}$).

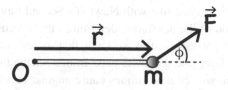

Figure 3.1 System of a mass m attached to the end of a massless rod. The rod is fixed at one end (i.e., point O) in such a way that the mass can only rotate about point O in the plane of the page, with the axis of rotation perpendicular to the page and passing through O. If the mass rotates, it will transcribe a circle with its center at O. A force $\vec{F}$ is exerted at the other end of the rod with $\vec{r}$ being the vector from the axis through point O to the point where the force is exerted.

As discussed in Chapter 2, since we know that only the component of $\vec{F}$ that is perpendicular to $\vec{r}$ causes mass m to start rotating, we can apply Newton's Second Law to mass m and see how that force component affects it:

$$\Sigma \vec{F} = m\vec{a} \Rightarrow |\vec{F}| \sin \phi = m|\vec{a}|. \tag{3.2}$$

Since a rotation will occur, let's rewrite the acceleration in terms of the rotational quantities we studied in Chapter 1. Since $a = r\alpha$, where α is the angular acceleration, we then see that

$$|\vec{F}| \sin \phi = m|\vec{a}| = m|\vec{r}||\vec{\alpha}|. \tag{3.3}$$

Multiplying both sides of this equation by $|\vec{r}|$, we find that the left side simply becomes the magnitude of the torque due to $\vec{F}$. The right side becomes a product of the constant $m|\vec{r}||\vec{r}| = mr^2$, which we will call the moment of inertia I of this point mass about an axis (with the SI unit of [kgm^2]), and the angular acceleration α:

$$|\vec{F}| \sin \phi = m|\vec{r}||\vec{\alpha}|$$

$$\Rightarrow \underbrace{|\vec{r}||\vec{F}| \sin \phi}_{|\vec{\tau}|} = \underbrace{m|\vec{r}|^2}_{mr^2 = I} \alpha$$

$$\Rightarrow |\vec{\tau}| = I|\vec{\alpha}|. \tag{3.4}$$

In the case of multiple forces, there could in principle be multiple torques on the mass about the same axis of rotation. We would then have to add all of the torques together to get the total angular acceleration. Thus, for rotational motion, we have a similar expression to Equation (3.1):

$$\Sigma \vec{\tau} = I\vec{\alpha}, \tag{3.5}$$

for a constant I. Recovering the vector Equation (3.5) from the magnitude Equation (3.4) is straightforward, since in this case the moment of inertia I

is a positive scalar.[1] So, just like with Newton's Second Law for translational motion, the direction of the net torque determines the direction of the angular acceleration. In other words, just as we know that forces cause linear accelerations with the net force and the linear acceleration having the same direction, in rotational motion we see that torques cause angular accelerations with the net torque and angular acceleration having the same direction. This is Newton's Second Law for rotational motion. A more mathematical way of obtaining Equation (3.5) is left as an exercise at the end of this chapter (a third way will also be given when we study angular momentum).

3.1.1 Notes on Newton's Second Law for Rotational Motion

(i) Though we will spend a whole chapter on the concept of the moment of inertia I, let's briefly discuss it here. In Equation (3.5) I plays the role of m in Equation (3.1). m is the constant of proportionality between the net force and the acceleration. Given the same net force on two different masses, the smaller mass will have the greater acceleration. Alternatively, for two different masses to obtain the same acceleration, the larger mass will require a larger net force. So, m indicates how hard it is to accelerate something. I shows the same thing! If we apply the same net torque to two different objects, the one with the smaller moment of inertia will rotate with the greater angular acceleration. Alternatively, if we wish to have the two objects rotate with the same angular acceleration, the object with the larger I will require a greater net torque and, thus, will be more difficult to rotate.

(ii) The above derivation found that $I = mr^2$, but this equation for I is true for a point mass only. We see that I depends not just on the mass but also on the distance r between the mass and the axis of rotation. Objects with greater mass require a greater torque to obtain the same angular acceleration if they have the same r. Alternatively, objects of the same mass will require different torques to obtain the same angular acceleration if they are located at different distances away from the axis of rotation. The mass with the greater distance from the axis of rotation will require more torque to achieve the same angular acceleration. In the next chapter, we will see that different objects have different moments of

---

[1] In reality, the moment of inertia I is a tensor (we will talk about this briefly in Chapter 8). When considering rotation about a fixed axis, which is always the case in this book, the moment of inertia *about this axis* is a scalar quantity. Though we do not expect the reader to be familiar with tensors, we want to make the point that treating the moment of inertia as a scalar is a special case.

inertia. Nevertheless, in all cases, $I > 0$ always and I depends both on the mass of the object and on how this mass is distributed about the axis of rotation.

(iii) Recall that $\Sigma \vec{F} = m\vec{a}$ was derived from the most general expression of Newton's Second Law, $\Sigma \vec{F} = \frac{d\vec{p}}{dt}$, where $\vec{p}$ is the linear momentum. $\Sigma \vec{F} = m\vec{a}$ is true only if m is constant. Similarly, $\Sigma \vec{\tau} = I\vec{\alpha}$ can be derived from $\Sigma \vec{\tau} = \frac{d\vec{L}}{dt}$ for constant I, where $\vec{L}$ is the angular momentum. We will see this derivation when we discuss angular momentum in detail in Chapter 5. The key point here is that Equation (3.5) is true only if I is constant, just like Equation (3.1) is true only if m is constant.

(iv) Since $I > 0$ always and the axis of rotation is fixed, $\Sigma \vec{\tau}$ and $\vec{\alpha}$ will always have the same direction along the fixed axis of rotation. This is analogous to the linear case: Since $m > 0$, $\Sigma \vec{F}$ and $\vec{a}$ always have the same direction in Newton's Second Law for translational motion.

(v) If I is constant, since it is a positive scalar, based on Equation (3.5):

(a) If
$$\Sigma \vec{\tau} = 0 \Leftrightarrow \vec{\alpha} = 0,$$
which means that the object is either stationary or moving with constant angular velocity ω.

(b) If
$$\Sigma \vec{\tau} = \text{constant} \neq 0 \Leftrightarrow \vec{\alpha} = \text{constant} \neq 0,$$
then we have rotational motion with constant $\vec{\alpha}$.

(c) If the net torque is not constant but is a function of position then, since I is constant, $\vec{\alpha}$ is not constant and is also a function of position.

(d) If the net torque is not constant but is a function of time then, since I is constant, $\vec{\alpha}$ is not constant and is also a function of time.

We looked at case (a) when we studied equilibrium in Chapter 2. Now we will focus on case (b). In cases (a) and (b), Equation (3.5) can thus be combined with the kinematics equations for rotational motion from Chapter 1. The methodology used is the same as when we used Newton's Second Law (Equation (3.1)) combined with the kinematics equations in translational motion. Cases (c) and (d) can be studied more easily using the Work-Kinetic Energy Theorem and the Momentum-Impulse Theorem as they apply to rotational motion, which we will study in later chapters.

Example 1: Torque on a horizontal disk. A homogeneous horizontal disk of mass $M = 2.0$ kg and $R = 0.5$ m is initially at rest and can rotate (no friction) on a smooth surface about the z axis that goes through its center and is

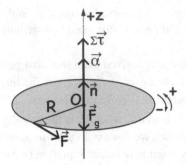

Figure 3.2 A homogeneous horizontal disk can rotate about the z axis that goes through its center and is perpendicular to the plane of the disk. The forces exerted on the disk are the gravitational force $\vec{F}_g$, the reaction force from the smooth surface $\vec{n}$, and the external force $\vec{F}$.

perpendicular to the plane of the disk, as shown in Figure 3.2. At time $t_0 = 0$ we exert a force $F = 100$ N on the disk at a point on the rim. The force is tangent to the disk's circumference and causes it to rotate counterclockwise as seen from above. Find (a) the disk's angular acceleration $\vec{\alpha}$, (b) the centripetal acceleration $\vec{a}_c$ of a point on the rim at $t_1 = 2.0$ s, and (c) the number of complete revolutions of the disk from $t_0 = 0$ to $t_1 = 2.0$ s. You are given that the moment of inertia of the disk about this axis is $I_z = 0.25$ kgm^2.

(a) We apply Newton's Second Law for rotational motion on the rotating disk, noting that the subscript z is used for $\Sigma\vec{\tau}$ to indicate the axis of rotation, as explained in Chapter 2. For the three forces shown in Figure 3.2, we get:

$$\Sigma\vec{\tau}_z = \underbrace{\vec{\tau}_{Fg}}_{0} + \underbrace{\vec{\tau}_n}_{0} + \vec{\tau}_F = I\vec{\alpha}. \tag{3.6}$$

Forces $\vec{F}_g$ and $\vec{n}$ do not produce torque about the z axis, since each force acts at this axis. Therefore, only the torque of $\vec{F}$ remains. This torque is the net torque and it is positive, since it causes counterclockwise rotation. Thus, it will produce a positive angular acceleration (i.e., both $\vec{\alpha}$ and $\Sigma\vec{\tau}$ point along the $+z$ axis). We then have

$$\tau_F = I_z\alpha$$

$$\Rightarrow RF \sin\left(\frac{\pi}{2}\right) = I_z\alpha$$

$$\Rightarrow \alpha = \frac{RF}{I_z} = \frac{0.5 \text{ m} \cdot 100 \text{ N}}{0.25 \text{ kgm}^2} = 200 \text{ rad/s}^2. \tag{3.7}$$

(b) From the definition of torque $\Sigma\tau = r\Sigma F \sin\phi$, since ΣF, r, and ϕ are constant, $\Sigma\tau$ is constant. Furthermore, the direction of $\Sigma\vec{\tau}$ does not change, as can be seen from the right-hand rule. Thus, $\vec{\alpha}$ is also constant and we have rotational motion with constant $\vec{\alpha}$. The equations that describe this concept are (see Chapter 1):

$$\Delta\theta = \omega_0\Delta t + \frac{1}{2}\alpha\Delta t^2, \tag{3.8}$$

and

$$\omega = \omega_0 + \alpha\Delta t. \tag{3.9}$$

Since all vectors in this problem are positive and along the z axis, we have written these equations in terms of magnitudes only. The disk is at rest initially, and so $\omega_0 = 0$. We have from Equation (3.9) for $t_1 = 2.0$ s:

$$\omega_1 = \alpha\Delta t = \alpha(t_1 - t_0) = 200 \text{ rad/s}^2 \cdot 2.0 \text{ s} = 400 \text{ rad/s}. \tag{3.10}$$

At that moment, therefore, v_1 at the rim is

$$v_1 = \omega_1 R = 400 \text{ rad/s} \cdot 0.5 \text{ m} = 200 \text{ m/s}. \tag{3.11}$$

Consequently, we get for the centripetal acceleration

$$a_c = \frac{v_1^2}{R} = \frac{(200 \text{ m/s})^2}{0.5 \text{ m}} = 8 \times 10^4 \text{ m/s}^2. \tag{3.12}$$

(c) When $t_1 = 2.0$ s, we can find $\Delta\theta$ from Equation (3.8) for $\omega_0 = 0$:

$$\Delta\theta_2 = \frac{1}{2}\alpha\Delta t^2 = \frac{1}{2}200 \text{ rad/s}^2(2.0 \text{ s})^2 = 400 \text{ rad}. \tag{3.13}$$

So, for the number of complete revolutions we have:

$$N = \frac{\Delta\theta}{2\pi} = \frac{400 \text{ rad}}{2\pi \text{ rad}} = \frac{200}{\pi} = 63.7 \text{ revolutions}. \tag{3.14}$$

This means the disk has completed 63 full revolutions.

Example 2: Mass on a rotating horizontal disk. A homogeneous disk of mass $M = 4.0$ kg and $R = 2.0$ m is on a smooth horizontal surface. It is rotating with constant angular velocity of magnitude $\omega_0 = 1.0$ rad/s about an axis z that goes through the disk's center and is perpendicular to the plane of the disk, as shown in Figure 3.3. A point mass $m = 1.0$ kg is placed on the disk at a distance $r = 1.0$ m from its center, as shown. The maximum coefficient of static friction between m and the disk is $\mu_s = 0.8$. At time $t_0 = 0$ we start exerting two forces on the disk at two points on the rim. The forces are tangent to the rim, have opposite directions, and are diametrically across from each other. The constant torque resulting from these two forces is in the same direction as that of $\vec{\omega}_0$. We find that the resulting angular acceleration has magnitude $\alpha = 2.0$ rad/s^2. Find (a) the magnitude of each force and (b) the time t when m is about to start

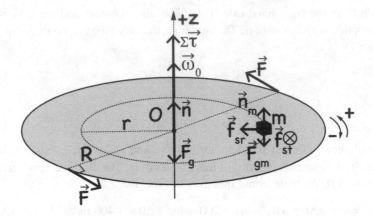

Figure 3.3 A homogeneous horizontal disk of mass M and radius R is rotating (no friction) about an axis z that goes through the disk's center and is perpendicular to the plane of the disk, due to a pair of forces. The static friction $\vec{f}_s$ exerted on m has two components, one toward the center of the disk, $\vec{f}_{sr}$ (which plays the role of the centripetal force), and the other tangent to the circle of radius r at the point where m is, $\vec{f}_{st}$.

sliding. You are given that the gravitational acceleration is $g = 10 \text{ m/s}^2$ and that the moment of inertia of the disk about this axis of rotation is $I_z = 8.0$ kgm^2.

(a) First, let's draw the forces exerted on the two objects. For the mass m, we have the gravitational force $\vec{F}_{gm}$, the normal force from the disk $\vec{n}_m$, and the static friction $\vec{f}_s$ between m and the disk. We have separated $\vec{f}_s$ into two components, $\vec{f}_{sr}$ in the radial direction, and $\vec{f}_{st}$ in the tangential direction. $\vec{f}_{sr}$ plays the role of the centripetal force for m, while $\vec{f}_{st}$ prevents m from slipping tangentially as the disk rotates.

For clarity, we have drawn only the following forces on the disk: The two forces $\vec{F}$, the gravitational force $\vec{F}_g$, and the normal force $\vec{n}$ on the disk by the smooth horizontal surface. We have not drawn the normal force from m on the disk ($\vec{n}'_m$) or the static friction force ($\vec{f}'_s = \vec{f}'_{sr} + \vec{f}'_{st}$) from m on the disk. However, the directions of these forces can be found using Newton's Third Law, since we have action–reaction pairs: $\vec{n}_m = -\vec{n}'_m$, $\vec{f}_{sr} = -\vec{f}'_{sr}$, and $\vec{f}_{st} = -\vec{f}'_{st}$.

Since m does not slide, both the disk and the mass m rotate together about the same axis of rotation z with the same angular acceleration (i.e., $\vec{\alpha}_{\text{disk}} = \vec{\alpha}_m = \vec{\alpha}$). We will apply Newton's Second Law for rotational motion for each mass separately:

For m:

$$\Sigma \vec{\tau}_z = \vec{\tau}_{Fgm} + \vec{\tau}_{n_m} + \vec{\tau}_{fsr} + \vec{\tau}_{fst} = I_m \vec{\alpha}. \tag{3.15}$$

For the disk:

$$\Sigma \vec{\tau}_z = \vec{\tau}_{Fg} + \vec{\tau}_n + \vec{\tau}_{n'_m} + \vec{\tau}_{f'sr} + \vec{\tau}_{f'st} + \vec{\tau}_F + \vec{\tau}_F = I_{\text{disk}} \vec{\alpha}. \tag{3.16}$$

If we add Equations (3.15) and (3.16) together, we see that the torques due to the action–reaction pairs are opposite and they will cancel out. That is, $\vec{\tau}_{n_m} = -\vec{\tau}_{n'_m}$, $\vec{\tau}_{fsr} = -\vec{\tau}_{f'sr}$, and $\vec{\tau}_{fst} = -\vec{\tau}_{f'st}$. We are left with:

$$\Sigma \vec{\tau}_{z.tot} = \vec{\tau}_{Fgm} + \vec{\tau}_{Fg} + \vec{\tau}_n + \vec{\tau}_F + \vec{\tau}_F = I_m \vec{\alpha} + I_{\text{disk}} \vec{\alpha}. \tag{3.17}$$

But $\vec{F}_{gm}$, $\vec{F}_g$, and $\vec{n}$ do not produce torque about the z axis since the forces are parallel to and/or along this axis. Therefore, only the torques due to the forces $\vec{F}$ remain and those torques are positive based on the right-hand rule and our choice of coordinate system. Using these conclusions, and that the moment of inertia of a point mass about an axis of rotation is $I = mr^2$, we have

$$\Sigma \vec{\tau}_{\text{z.tot}} = I_{\text{disk}} \vec{\alpha} + I_m \vec{\alpha}$$

$$\Rightarrow 2\tau_F = (I_{\text{disk}} + mr^2)\alpha$$

$$\Rightarrow 2FR \sin\left(\frac{\pi}{2}\right) = (I_{\text{disk}} + mr^2)\alpha$$

$$\Rightarrow 2F(2.0 \text{ m}) = (8.0 \text{ kgm}^2 + 1 \text{ kg} \cdot (1.0 \text{ m})^2) \cdot 2.0 \text{ rad/s}^2$$

$$\Rightarrow F = 4.5 \text{ N}. \tag{3.18}$$

(b) Now our focus is on the mass m, since we are interested in knowing the time t when it will start slipping. Since we have:

$$\vec{f}_s = \vec{f}_{st} + \vec{f}_{sr}, \tag{3.19}$$

and we know that $\vec{f}_{st}$ is perpendicular to $\vec{f}_{sr}$, then

$$f_s = \sqrt{f_{st}^2 + f_{sr}^2}. \tag{3.20}$$

From Newton's Second Law, we have (using $a_t = \alpha r$, $v = \omega r$, and $a_r = a_{\text{centripetal}}$):

$$\Sigma \vec{F}_t = m\vec{a}_t$$

$$\Rightarrow f_{st} = ma_t = m\alpha r, \tag{3.21}$$

and

$$\Sigma \vec{F}_r = m\vec{a}_r$$

$$\Rightarrow f_{sr} = \frac{mv^2}{r} = m\omega^2 r. \tag{3.22}$$

Thus,

$$f_s = \sqrt{f_{st}^2 + f_{sr}^2} = \sqrt{(m\alpha r)^2 + (m\omega^2 r)^2}. \tag{3.23}$$

Finally, at the instant when m is about to start sliding, $f_s = f_{s\max}$ with $f_{s\max} = \mu_s n \Rightarrow f_{s\max} = \mu_s mg$ ($n = mg$ because m is in equilibrium along the z axis). Combining this with Equation (3.23) we get:

$$\mu_s mg = \sqrt{(m\alpha r)^2 + (m\omega^2 r)^2}$$

$$\Rightarrow (\mu_s g)^2 = (\alpha r)^2 + \omega^4 r^2$$

$$\Rightarrow \omega = \left[\frac{(\mu_s g)^2 - (\alpha r)^2}{r^2} \right]^{1/4}$$

$$\Rightarrow \omega = \left[\frac{\left(0.8 \cdot 10 \text{ m/s}^2\right)^2 - (2.0 \text{ rad/s}^2 \cdot 1.0 \text{ m})^2}{(1.0 \text{ m})^2} \right]^{1/4} = 2.8 \text{ rad/s}. \tag{3.24}$$

Now we know ω at that time t. To find t we can use the equations for rotational motion with constant angular acceleration

$$\Delta\theta = \omega_0 \Delta t + \frac{1}{2}\alpha\Delta t^2, \tag{3.25}$$

$$\omega = \omega_0 + \alpha\Delta t. \tag{3.26}$$

Again, since all vectors in this problem are positive and along the z axis, we have written these equations in terms of magnitudes only. From Equation (3.26) we have:

$$2.8 \text{ rad/s} = 1.0 \text{ rad/s} + 2.0 \text{ rad/s}^2 (t - 0) \Rightarrow t = 0.9 \text{ s}. \tag{3.27}$$

Example 3: A swinging rod-mass system. A thin uniform rod OA has mass $M = 1.0$ kg and length $L = 1.0$ m. A small point mass $m = 0.1$ kg is firmly attached to the rod at its end A. The rod's other end O is attached to a vertical wall via a junction. The junction allows the rod to rotate (no friction) about a horizontal axis that goes through point O and is perpendicular to the rod and the plane of the page, as shown in Figure 3.4a. At time $t_0 = 0$ we release the rod from rest from its initial horizontal position and it starts rotating. (a) Write the equations that relate the magnitude of the rod-mass system's angular acceleration to the angle β the rod forms with its initial position at any instant t. Then sketch the graph for $\beta \in [0, \frac{\pi}{2}]$. (b) Explain why $R_x = 0$ at time $t_0 = 0$ s, but $R_x \neq 0$ at any other time $t > 0$ while the rod-mass system is falling. (c) Find the tangential acceleration of the rod's midpoint B at the instant when the rod is vertical, right before it hits the wall. (d) If at time $t_0 = 0$ (when we let the rod go) we placed a box on top of the rod at midpoint B, would the box and the rod separate as the rod starts rotating? Justify your answer. You are given

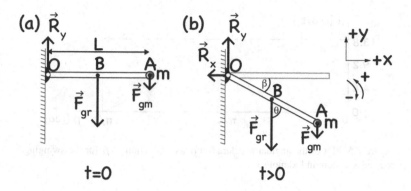

Figure 3.4 (a) A thin uniform rod OA has a mass m attached to one end while the other end is attached to the wall via a junction. The rod can rotate about a horizontal axis that goes through point O and is perpendicular to the rod and the plane of this page. At time $t_0 = 0$ the rod is released from its initial horizontal position. (b) All the forces exerted at some arbitrary time $t > 0$ while the rod-mass system is swinging, and the rod forms an angle β with its initial, horizontal position.

$g = 10$ m/s^2 and that the moment of inertia of this rod-mass system about an axis of rotation that goes through the rod's end O is $I_{system} = \frac{13}{30}$ kgm^2.

(a) As always, we draw all the forces exerted on the rod at some arbitrary time $t > 0$ and we show our coordinate system, as seen in Figure 3.4b. The reaction force from the junction ($\vec{R}$) is decomposed into its components $\vec{R}_x$ and $\vec{R}_y$, where the directions will be explained shortly. We also have the gravitational forces $\vec{F}_{gr}$ on the rod and $\vec{F}_{gm}$ on the mass. Both $\vec{R}_x$ and $\vec{R}_y$ are exerted at the axis of rotation (i.e., $\vec{r} = 0$) and thus they do not produce torque about this axis.

The rod and mass m are unable to move independently of each other – they are stuck together so we can treat them as one object. Using Newton's Second Law for rotational motion, we have:

$$\Sigma \vec{\tau}_O = I_{system} \vec{\alpha}$$

$$\Rightarrow \underbrace{\tau_{Rx}}_{0} + \underbrace{\tau_{Ry}}_{0} - \tau_{Fgr} - \tau_{Fgm} = -I_{system}\alpha$$

$$\Rightarrow F_{gr}\frac{L}{2}\sin\theta + F_{gm}L\sin\theta = I_{system}\alpha$$

$$\Rightarrow Mg\frac{L}{2}\cos\beta + mgL\cos\beta = I_{system}\alpha$$

$$\Rightarrow \alpha = \frac{Lg\cos\beta}{I_{system}}\left(\frac{M}{2} + m\right) = \frac{1.0\,\text{m}\cdot\left(10\,\text{m/s}^2\right)\cos\beta}{\frac{13}{30}\,\text{kgm}^2}\left(\frac{1.0\,\text{kg}}{2} + 0.1\,\text{kg}\right)$$

$$\Rightarrow \alpha = 13.8\cos\beta\,(\text{rad/s}^2). \qquad (3.28)$$

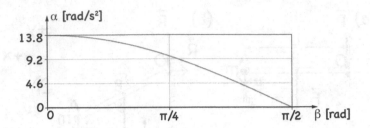

Figure 3.5 Plot of the angular acceleration α as a function of β for the swinging rod-mass system in Example 3.

In this calculation, we have used the fact that $\beta + \theta = \pi/2$, as seen in Figure 3.4b. Thus, $\sin \theta = \cos \beta$. The plot of the angular acceleration α as a function of β is shown in Figure 3.5.

(b) From Figure 3.4b we see that $\vec{R}_x$ is the only force along the x axis. At time $t_0 = 0$ s, therefore, $\vec{R}_x$ is the only force in the radial direction that can play the role of the centripetal force; all the other forces are along the y axis and have a tangential direction to the rod-mass system's circular trajectory. From Newton's Second Law for the centripetal force we have:

$$\Sigma F_c = ma_c = m\frac{v^2}{r}$$

$$\Rightarrow R_x = m\frac{v^2}{r}. \tag{3.29}$$

At time $t_0 = 0$ s the rod is released from rest. Thus, $v_0 = 0$. Consequently, $R_x(t_0) = 0$. At any other time $t > 0$, due to the system's acceleration, $v(t > 0) \neq 0$ for the rod-mass system. Therefore, a component of $\vec{R}$ in the *radial* direction must play the role of the centripetal force. A nonzero radial component of $\vec{R}$ implies that $R_x(t > 0) \neq 0$.

(c) When the rod is vertical, $\beta = \pi/2$. Therefore, from our result in part (a) we have:

$$\alpha = 13.8 \cos \left(\frac{\pi}{2}\right) = 0. \tag{3.30}$$

Then, since $a = \alpha r$ we have:

$$a_B = \alpha \frac{L}{2} = 0. \tag{3.31}$$

This result makes sense: When the rod-mass system is vertical, the tangential direction is along the x axis and the radial direction is along the y axis. At that instant, there are no forces in the tangential direction. Therefore, the tangential acceleration at any point along the rod, including point B, is zero.

(d) The moment when the rod is released ($t_0 = 0$ s), the acceleration of the rod at point B is

$$a_B = \alpha r = \left(\left(13.8 \text{ rad/s}^2\right) \cos 0\right)\frac{L}{2} = \left(13.8 \text{ rad/s}^2\right)\frac{L}{2} = \left(13.8 \text{ rad/s}^2\right)\frac{1.0 \text{ m}}{2}$$

$$\Rightarrow a_B = 6.9 \text{ m/s}^2. \tag{3.32}$$

We see that $a_B < g$, so the box and the rod will remain together. In fact, when the box is initially in contact with the rod, the reaction force from the box on the rod will push down on the rod, so the rod's acceleration (of its midpoint) will slightly increase. If $a_B > g$, then the rod and the box would be separated, since the box would be accelerating at g and the rod at $a_B > g$.

3.2 Systems of Objects Combining Translational and Rotational Motion

Now we will look at a system of objects wherein some objects execute translational motion and the rest execute rotational motion. As always, when we have a system of objects that can move independently of each other, we look at each object in the system separately and then we look for clues in the problem that provide connections between the objects. Since problems like this contain both translational and rotational motion, it is certain that we will have to use the equations $v = \omega r$ and/or $a = \alpha r$. These equations provide the connection between the translational parameters v and a to the rotational parameters ω and α. Let's do an example, where the methodology is clearly shown.

Example 4: Mass-pulley system. A uniform pulley of radius r, mass M, and moment of inertia I is free to rotate (no friction) about a horizontal axle that goes through its center O and is perpendicular to the plane of the pulley. A massless and inelastic rope is wound around the pulley at its circumference and a mass m is hanging from the rope's free end. This system is shown in Figure 3.6. At time t_0, the system is released from rest. Find (a) the angular acceleration of the pulley, (b) the linear acceleration of the mass m, (c) the tension on the rope, (d) the force $\vec{R}$ exerted by the axle on the pulley, (e) the centripetal acceleration of a point on the rim of the pulley at time t, and (f) the displacement of m between t_0 and any time t. Assume that the rope is not slipping against the rim of the pulley and that g is given.

As always, we draw the picture with all the forces clearly shown and labeled, and we indicate the coordinate system, as seen in Figure 3.6. We will study each

3 *Rotational Dynamics*

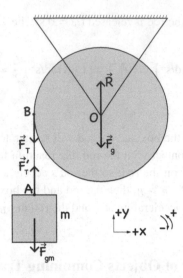

Figure 3.6 A uniform pulley is free to rotate about a horizontal axle that goes through its center O and is perpendicular to the plane of the pulley. A massless and inelastic rope is wound around the pulley at its circumference and a mass is hanging from the rope's free end.

mass separately and then look for clues that connect the two masses together. There are three forces exerted on the pulley: The gravitational force $\vec{F}_g$, the reaction force $\vec{R}$ from the axle, and the force due to tension from the rope $\vec{F}_T$. There are only two forces exerted on the mass m: The gravitational force $\vec{F}_{gm}$ and the force due to tension from the rope $\vec{F}_T'$.

For the pulley M:
x axis: There are no forces along this axis, so we have equilibrium $\Sigma \vec{F}_x = 0$.
y axis: The pulley is not executing linear motion along this axis. So, the pulley is again in equilibrium along the y axis, and we have:

$$\Sigma \vec{F}_y = 0 \Rightarrow R - F_g - F_T = 0 \Rightarrow R = F_g + F_T. \qquad (3.33)$$

However, the pulley does execute rotational motion when m starts moving downwards. So, we apply Newton's Second Law for rotational motion. The subscript O of the net torque indicates that the axis of rotation goes through point O and is perpendicular to the plane of the page.

$$\Sigma \vec{\tau}_O = I\vec{\alpha}$$

$$\Rightarrow \underbrace{\tau_R}_{0} + \underbrace{\tau_{Fg}}_{0} + \tau_{F_T} = I\alpha$$

$$\Rightarrow F_T r \sin\left(\frac{\pi}{2}\right) = I\alpha$$

$$\Rightarrow \alpha = \frac{F_T r}{I}, \tag{3.34}$$

where we have used that the torques due to $\vec{F}_g$ and $\vec{R}$ are zero, since both forces are exerted at the axis of rotation. Therefore, only the torque of the force due to tension remains. This force is constant and the axis of rotation does not change. Thus, $\Sigma \vec{\tau}_O$ is constant and so we have rotational motion with constant angular acceleration $\vec{\alpha}$. The equations that describe this concept are (with $\omega_0 = 0$ since the pulley starts from rest):

$$\Delta\theta = \underbrace{\omega_0}_{0}\Delta t + \frac{1}{2}\alpha\Delta t^2 = \frac{1}{2}\alpha\Delta t^2, \tag{3.35}$$

$$\omega = \underbrace{\omega_0}_{0} + \alpha\Delta t = \alpha\Delta t. \tag{3.36}$$

We have written these equations in terms of magnitudes only because the torque of the force due to tension will produce counterclockwise (positive) rotation. Since we do not know F_T, we cannot yet find R or α based on the above equations. So, we move on to study m and we will come back to M shortly.

For the mass m:

x axis: There are no forces along this axis, so we have equilibrium $\Sigma\vec{F}_x = 0$.

y axis: The mass has a translational acceleration along this axis, so we apply Newton's Second Law:

$$\Sigma\vec{F}_y = m\vec{a}$$

$$\Rightarrow F_T' - F_{gm} = -ma$$

$$\Rightarrow a = \frac{mg - F_T'}{m}. \tag{3.37}$$

Since $\Sigma\vec{F}_y$ is constant, $\vec{a}$ is constant. Therefore, the hanging mass executes $1D$ motion with constant acceleration and the equations that describe this motion are (with $v_0 = 0$ since the mass starts from rest, and the correct algebraic signs based on our coordinate system):

$$\Delta y = \underbrace{v_0}_{0} \Delta t - \frac{1}{2} a \Delta t^2 = -\frac{1}{2} a \Delta t^2, \tag{3.38}$$

$$v_y = \underbrace{v_0}_{0} -a\Delta t = -a\Delta t. \tag{3.39}$$

We are now done with all the physical concepts and equations. Let's look for clues that connect M and m together.

(i) Since the rope is inelastic, from Newton's Third Law $F_T = F_T'$.
(ii) Since the rope is inelastic, massless, and is not slipping against the pulley's rim, all points on the rope have the same linear velocity and acceleration. So, the point on the rope in contact with the rim of the pulley (point B) and the point on the rope in contact with mass m (point A) have the same linear velocity and acceleration. Thus, points on the rim of the pulley and the mass m also have the same linear velocity and acceleration:

$$v_A = v_B = \omega r \tag{3.40}$$
$$a_A = a_B = \alpha r. \tag{3.41}$$

Now, let's plug Equations (3.34), (3.37), and the equation from the clue that $F_T = F_T'$ into the above equation $a = \alpha r$:

$$a = \alpha r$$

$$\Rightarrow \frac{mg - F_T'}{m} = \frac{F_T r}{I} r$$

$$\Rightarrow mg - F_T = \frac{m F_T r^2}{I}$$

$$\Rightarrow mg = F_T \left(\frac{mr^2}{I} + 1 \right)$$

$$\Rightarrow F_T = \frac{mg}{\frac{mr^2}{I} + 1}. \tag{3.42}$$

This is the answer to part (c).

(a) Now, we can plug this result into Equation (3.34)

$$\alpha = \frac{F_T r}{I} = \frac{mg}{\frac{mr^2}{I} + 1} \cdot \frac{r}{I} = \frac{mgr}{mr^2 + I}, \tag{3.43}$$

which is the pulley's angular acceleration.

(b) Since $a = \alpha r$, then we can use our result from part (a) for α to find a:

$$a = \alpha r = \frac{mgr^2}{mr^2 + I} \tag{3.44}$$

is the acceleration of the mass m.

(d) Now we can plug the result of part (c) in Equation (3.33) to find R:

$$R = F_g + F_T = Mg + F_T = g\left(M + \frac{m}{\frac{mr^2}{I} + 1}\right). \tag{3.45}$$

(e) At time t, Equation (3.39) gives

$$v_y = -a\Delta t = -\frac{mgr^2\Delta t}{mr^2 + I}. \tag{3.46}$$

As we discussed, the points on the rim have the same linear speed as m. So, for the centripetal acceleration a_c we have ($v_y = v$):

$$a_c = \frac{v^2}{r} = \left(\frac{mgr^2\Delta t}{mr^2 + I}\right)^2 \frac{1}{r} - \left(\frac{mg\Delta t}{mr^2 + I}\right)^2 r^3, \tag{3.47}$$

where $\Delta t = t - t_0$.

(f) From Equation (3.38), if we use our result from part (b), we get:

$$\Delta y = -\frac{1}{2}a\Delta t^2 = -\frac{1}{2}\frac{mgr^2}{mr^2 + I}\Delta t^2. \tag{3.48}$$

Exercises

(i) Derive Newton's Second Law for rotational motion (Equation (3.5)) in another way: Start from Newton's Second Law $\Sigma\vec{F} = m\vec{a}$, take the cross product $\vec{r} \times \Sigma\vec{F}$, and use the result of Exercise 5 in Chapter 1 that $\vec{a}_t = \vec{\alpha} \times \vec{r}$. You might need to use the identity that $\vec{A} \times (\vec{B} \times \vec{C}) = \vec{B}(\vec{A} \cdot \vec{C}) - \vec{C}(\vec{A} \cdot \vec{B})$.

(ii) Four point masses are stuck on the surface of a disk, at the rim, as shown in Figure 3.7. The whole system is vertical and is free to rotate about an axis that passes through its center O and is perpendicular to the plane of the page. At time $t_0 = 0$ the system is released from rest. Find the system's angular acceleration just after it is released, given that $g = 10$ m/s^2, the radius of the disk is $R = 2.5$ m, and that the system's moment of inertia about the axis of rotation through O is $I = 422$ kgm^2.

(iii) Shown in Figure 3.8 is a pulley consisting of two concentric, nonuniform disks of radii $R = 1.0$ m and $r = 0.5$ m which are stuck together. The pulley can rotate about a fixed axis that goes through its center and is perpendicular to the plane of the page. Two separate pieces of a massless, inelastic string are wrapped around the circumferences of the disks, and their free ends are attached to masses $m_1 = 3.0$ kg and $m_2 = 5.0$ kg. The whole system is initially at rest, and though the pulley is fixed, the masses can slide without friction on the smooth horizontal surface. At

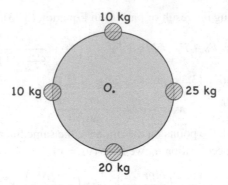

Figure 3.7 Four point masses are stuck on the surface of a vertical disk at the rim. Once released from rest, the system will rotate about an axis that goes through its center O and is perpendicular to the plane of the page.

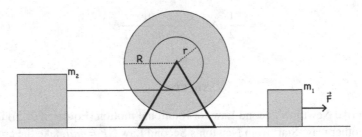

Figure 3.8 A pulley consists of two concentric disks and is free to rotate about an axis that goes through its center and is perpendicular to the plane of the page. Massless and inelastic strings are wound around the pulley at its inner and outer disks, and each string is attached to a mass. When a force $\vec{F}$ is exerted on mass m_1, the masses slide on the smooth horizontal surface and the pulley rotates.

time $t_0 = 0$ we start exerting a horizontal force $\vec{F}$ of magnitude 50 N pointing to the right, as shown, and the masses start to move as the pulley begins to rotate. During its motion, m_1 experiences a tension force from its string of magnitude $F_{T_1} = 15$ N. You are given that $g = 10 \text{ m/s}^2$ and that there is no slipping of the strings on the disks' circumferences. Find (a) the angular acceleration of the pulley, (b) the moment of inertia of the pulley about its axis of rotation, and (c) the magnitude of the torque due to the tension force exerted on the pulley by the string attached to m_2.

(iv) A uniform pulley of radius $R = 0.06$ m and mass $M = 2.0$ kg is mounted on an axle that goes through its center O. A massless and inelastic cord is wrapped around its rim, with a mass attached to each end of the cord ($m_1 = 2.0$ kg and $m_2 = 4.0$ kg), as shown in Figure 3.9. The difference in

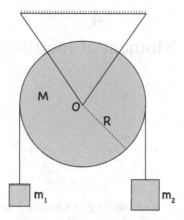

Figure 3.9 A uniform pulley of radius R and mass M is mounted on an axle that goes through its center O and is perpendicular to the plane of the page. A massless and inelastic cord is wrapped around its rim, with a mass attached to each end of the cord.

the two masses causes the pulley to rotate without slipping, but as it rotates, friction at the axle exerts a torque on the pulley of magnitude $\tau_f = 0.5$ Nm. The pulley's moment of inertia about the axle is $I = 0.0036$ kgm^2 and $g = 10$ m/s^2. If the system is released from rest, find (a) the linear acceleration of the hanging masses and (b) how long it takes m_1 to travel up 1 m. (c) Explain why the tensions at each end of the cord are not equal.

4

Moment of Inertia

We first defined the moment of inertia I in Chapter 3 when we derived Newton's Second Law for rotational motion: $\Sigma \vec{\tau} = I \vec{\alpha}$. We discussed that I is the rotational equivalent of m in $\Sigma \vec{F} = m \vec{a}$. For a given net torque, I determines the magnitude of an object's angular acceleration, just like for a given net force, m determines the magnitude of an object's linear acceleration. We had found that $I = mr^2$, but this equation for I is true for a point mass only. Nevertheless, $I > 0$ always. For objects other than point masses, I depends not just on the total mass but also on how the mass is distributed about the axis of rotation. In this chapter, we will start from the equation of the moment of inertia for a point mass and extend this definition so that we can find the moment of inertia of any rigid body.

4.1 Moment of Inertia of Systems of Discrete Masses and Continuous Rigid Bodies

Let's start with an arbitrarily shaped rigid body, for example, a potato, shown in Figure 4.1. Imagine that we apply a force that causes the potato to rotate with some angular acceleration about an axis z that remains fixed. We will first consider this solid object as a collection of n small masses m_1, m_2, ..., m_n, which are at distances $r_1, r_2, ..., r_n$, respectively, from the axis of rotation. Because it is a rigid body and the distance between any two small masses remains constant, it must be the case that every small mass experiences the effect of the applied force. Thus, each small mass will experience a torque $\vec{\tau}_1 = I_1 \vec{\alpha}_1$, $\vec{\tau}_2 = I_2 \vec{\alpha}_2$, ..., $\vec{\tau}_n = I_n \vec{\alpha}_n$, respectively, about the axis of rotation. Then, for the object as a whole, we will have for the net torque about the z axis:

$$\Sigma \vec{\tau}_z = \vec{\tau}_1 + \vec{\tau}_2 + ... + \vec{\tau}_n$$
$$= I_1 \vec{\alpha}_1 + I_2 \vec{\alpha}_2 + ... + I_n \vec{\alpha}_n$$
$$= m_1 r_1^2 \vec{\alpha}_1 + m_2 r_2^2 \vec{\alpha}_2 + ... + m_n r_n^2 \vec{\alpha}_n. \tag{4.1}$$

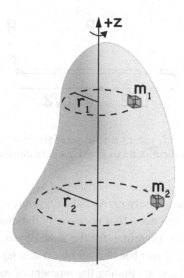

Figure 4.1 A potato of mass M represents any arbitrarily shaped rigid object. This object is rotating about an axis z that remains fixed. The potato consists of many small masses m_1, m_2, etc. that are at distances r_1, r_2, etc., respectively, from the axis of rotation.

Again, since this is a rigid body, all the small masses will rotate with the same angular acceleration $\vec{\alpha}_1 = \vec{\alpha}_2 = \ldots = \vec{\alpha}_n = \vec{\alpha}$. So, the above equation becomes:

$$\Sigma \vec{\tau} = \left(m_1 r_1^2 + m_2 r_2^2 + \ldots + m_n r_n^2\right) \vec{\alpha} = \left(\Sigma_{i=1}^{n} m_i r_i^2\right) \vec{\alpha}. \qquad (4.2)$$

From this equation, we see that the moment of inertia of this arbitrarily shaped extended rigid body is

$$I = \Sigma_{i=1}^{n} m_i r_i^2. \qquad (4.3)$$

Of course, this definition for the moment of inertia works very well when studying discrete small masses rotating about the same axis of rotation. However, the potato is really a continuous object. Therefore, we can make this sum more accurate by considering the limit where the masses become infinitesimal: dm_1, dm_2, ..., dm_n, so that the sum becomes an integral over m:

$$I = \int r^2 dm. \qquad (4.4)$$

Just like for a single point mass where $I = mr^2$, we again see from these definitions that the moment of inertia about a specified axis is a scalar quantity with units of [kgm^2]. Its value depends on the mass of the object (via the infinitesimal masses), the axis of rotation, and how the masses are distributed about that

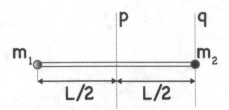

Figure 4.2 Two masses m_1 and m_2 are connected via a massless rod of length L. The moment of inertia about each axis p and q will be different.

axis (via the value of the quantity r). As before, r represents the distance from the axis of rotation to each mass (not necessarily the distance from the origin of the coordinate system). Therefore, for the same object, I can take an infinite number of values (all positive) because an object can have an infinite number of axes of rotation. Thus, when finding the moment of inertia about an axis, it is essential to specify about which axis we are evaluating it. Lastly, when evaluating the integral of Equation 4.4, we need to remember that r changes with each dm. Therefore, it is often necessary to use an object's density to relate dm and r, as we will show in Example 3.

In the following examples, we will calculate the value of the moment of inertia for a few different distributions of mass. Of course, there are many other mass distributions for which we don't perform these calculations in this book. Moments of inertia of many common, symmetric mass distributions (e.g., spheres, boxes) about various axes of rotation can be found in tables online and in other textbooks (or you can compute them yourself using the tools we learn here).

Example 1: The moment of inertia of a system of masses. The system shown in Figure 4.2 consists of two point masses $m_1 = 0.1$ kg and $m_2 = 0.2$ kg connected together via a rigid, massless rod of length $L = 0.2$ m. Find the system's moment of inertia about an axis that (a) goes through the midpoint of the rod (axis p) and (b) goes through the end where mass m_2 is located (axis q).

(a) Since we have a system of masses, we must account for the moment of inertia of each mass about the specified axis of rotation. As always, we use a subscript to indicate the axis of rotation about which we find I:

$$I_p = I_{m_1} + I_{m_2} + I_{\text{rod}}. \tag{4.5}$$

This equation is an extension of Equation (4.3) to include continuous rigid bodies. However, in this case, $I_{\text{rod}} = 0$ since the rod is massless ($M = 0$). Therefore, we obtain:

$$I_p = I_{m_1} + I_{m_2} = m_1 r_1^2 + m_2 r_2^2, \tag{4.6}$$

where we have used the fact that the two masses are point masses and, thus, their moment of inertia is given by $I = mr^2$. In this case, $r_1 = r_2 = L/2$. Therefore, we have:

$$I_p = m_1 \left(\frac{L}{2}\right)^2 + m_2 \left(\frac{L}{2}\right)^2 = 0.1 \text{ kg} \left(\frac{0.2 \text{ m}}{2}\right)^2 + 0.2 \text{ kg} \left(\frac{0.2\text{m}}{2}\right)^2 = 3.0 \times 10^{-3} \text{kgm}^2.$$
$$\tag{4.7}$$

(b) We will again start with Equation (4.3) using the subscript q to indicate the axis of rotation:

$$I_q = I_{m_1} + I_{m_2} + I_{\text{rod}}. \tag{4.8}$$

However, now both I_{m_2} and I_{rod} are zero. The former because the distance from m_2 to the axis of rotation q is zero and the latter because, just like before, the rod is massless. So, our equation is simply:

$$I_q = I_{m_1} = m_1 r_1^2. \tag{4.9}$$

In this scenario, the distance between m_1 and the axis of rotation is $r_1 = L$. So, we have:

$$I_q = m_1 L^2 = 0.1 \text{ kg} \cdot (0.2 \text{ m})^2 = 4.0 \times 10^{-3} \text{ kgm}^2. \tag{4.10}$$

Therefore, it is more difficult to rotate this system about axis q than p. In other words, the same net torque would produce a smaller angular acceleration about axis q.

Example 2: The moment of inertia of a uniform ring. Find the moment of inertia of a uniform circular ring of radius R and mass M about an axis of rotation that goes through the ring's center O and is perpendicular to its plane.

Since a ring is a continuous rigid object, we have to use Equation (4.4). Since m is the variable of integration whenever we have a continuous object, we divide it into infinitesimal mass elements dm, as shown in Figure 4.3. We

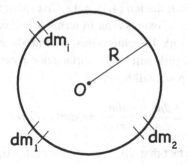

Figure 4.3 A uniform ring of radius R and mass M is divided into infinitesimal masses dm. Each dm is the same distance R away from the axis of rotation.

always examine at least two different (arbitrarily picked) elements (dm_1 and dm_2) to see if there is a pattern that will help us evaluate the integral. In this case we see that no matter which dm we pick (dm_1, dm_2, ..., dm_i), they are all the same distance $r = R$ from the axis of rotation. Therefore, $r = R$ is a constant and independent of m in this case. We then have:

$$I_O = \int_{\text{ring}} r^2 dm = \int_{\text{ring}} R^2 dm = R^2 \int_{\text{ring}} dm = MR^2, \qquad (4.11)$$

since the integral (sum) of dm over the ring is the total mass of the ring M, that is, $\int dm = M$. Now, let's find the moment of inertia of a uniform disk with the same mass M and radius R to see how they compare.

Example 3: The moment of inertia of a uniform disk. Find the moment of inertia of a thin, uniform disk of radius R and mass M about an axis of rotation that goes through the disk's center O and is perpendicular to its plane.

A disk is a continuous, rigid object so we again have to use Equation (4.4), keeping in mind that the variable of integration is m. Just like before, we divide the disk into infinitesimal mass elements dm. As shown in Figure 4.4a, in this case our dm elements correspond to circular strips, each of area dA. The reason why we pick strips is because we want to exploit the symmetry of the disk. Because each circular strip has radius r, all of the small masses that make up a given strip are the same distance r away from the axis of rotation (since the width of the strip dr is infinitesimal). In other words, each strip is a ring, and we are adding up these rings to create a disk.

Following the same methodology as in the previous example, let's examine two different mass elements dm_1 and dm_2 to see if there is a pattern that will help us evaluate the integral. dm_1 is a distance r_1 from the axis of rotation while dm_2 is a distance r_2 from the axis of rotation. For the rings that make up the disk, therefore, we see that r will vary from 0 to R. However, m is the variable of integration in Equation (4.4), not r. To exploit the fact that we know the limits of r, we need to express dm in terms of r (i.e., we need to change the variable of integration). To achieve this, we introduce the object's density, which in the case of the disk will be the surface density σ, defined as mass per area. Then for each strip we will have:

$$\sigma = \frac{dm}{dA} = \frac{dm}{2\pi r\, dr} \Rightarrow dm = \sigma 2\pi r\, dr. \qquad (4.12)$$

Here we have used the fact that $dA = 2\pi r\, dr$ for each strip. The reason for this is clear if we imagine that we take one of these strips from the disk, cut it along

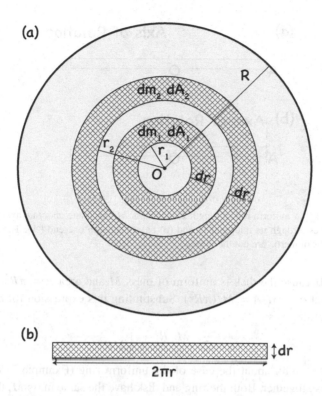

Figure 4.4 (a) A uniform disk of radius R and mass M is divided into circular strips, each of which has an infinitesimal mass dm, an infinitesimal width dr, an infinitesimal area dA, and is a distance r away from the axis of rotation. (b) We cut out a circular strip of radius r from the disk centered about point O. This strip corresponds to a mass dm which is part of the disk. Then we further cut the strip along dr and unroll it. We see that we have obtained a rectangle whose width is dr, length is $2\pi r$, and thus the area is $2\pi r\, dr$.

dr, and then unroll it. As shown in Figure 4.4b, the strip unrolls into a rectangle of length $2\pi r$ (the ring's circumference) and width dr. After all this, we see that we have effectively expressed dm in terms of dr, which means that it is now easy to evaluate the integral. We have:

$$I_O = \int r^2 dm = \int_0^R r^2 \sigma 2\pi r\, dr = \int_0^R r^3 \sigma 2\pi\, dr. \qquad (4.13)$$

Since the disk is uniform, σ is constant. Therefore, $2\pi\sigma$ is constant and it can come out of the integral:

$$I_O = 2\pi\sigma \int_0^R r^3 dr = 2\pi\sigma \frac{R^4}{4}. \qquad (4.14)$$

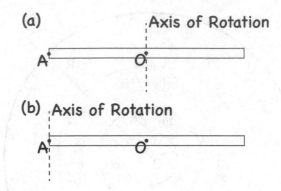

Figure 4.5 A uniform rod of length L and mass M can rotate about an axis that
(a) passes through its midpoint O and (b) passes through one end (A). The two
moments of inertia are not the same.

However, because the disk is uniform of mass M and area $A = \pi R^2$, we can
also say that $\sigma = M/A = M/(\pi R^2)$. Substituting this expression for σ in our
previous result gives:

$$I_O = 2\pi \frac{M}{\pi R^2} \frac{R^4}{4} = \frac{1}{2} M R^2. \tag{4.15}$$

Let's now think about the case of the uniform ring (Example 2) and this
uniform disk together. Both the ring and disk have the same mass M, the same
radius R, and an axis of rotation that passes through the object's center and is
perpendicular to its plane. Why is it then that the ring's moment of inertia is
larger? In the ring's case, its whole mass is the maximum possible distance R
away from the axis of rotation ($r = R$). On the other hand, in the disk's case,
some of its mass is closer to the axis of rotation ($r < R$) – only a portion of
its mass M is the maximum distance R away from its rotation axis. Therefore,
because of this difference in the mass distribution about the axis of rotation,
$I_{\text{ring}} > I_{\text{disk}}$.

Example 4: The moment of inertia of a rod about two different axes. Find
the moment of inertia of a uniform, rigid rod of length L and mass M about an
axis of rotation that is perpendicular to the rod and (a) passes through the rod's
center of mass (its midpoint) (Figure 4.5a) and (b) passes through one end of
the rod (Figure 4.5b).

This is an interesting example because it allows us to showcase one of the
subtleties mentioned earlier: r in the equation for the moment of inertia is the
distance from the axis of rotation and is always positive. r is not the position
with respect to the origin, which could be positive or negative. To start, let's

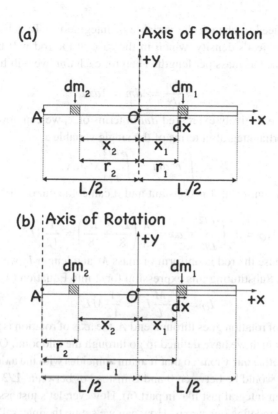

Figure 4.6 (a) When the y axis and the axis of rotation coincide, then $r = |x|$, with x between $-L/2$ and $L/2$ (b) When the axis of rotation for the uniform rod passes through point A but the y axis passes through the midpoint, then the distance of each infinitesimally small mass from the axis of rotation is not the same as the absolute value of the position of this mass along the x axis. r takes values between 0 and L but x is still between $-L/2$ and $L/2$.

imagine we place the rod along the x axis and we make the y axis pass through the midpoint O.

(a) In this case the y axis is the axis of rotation. If we divide the rod into infinitesimal masses dm, each of length dx, then for two arbitrarily picked dm elements, we see that $r_1 = x_1$ and $r_2 = |x_2|$, as shown in Figure 4.6a. In addition, $dr = dx$. So, we can express r in terms of x.

Once again, just like in the example with the disk, we see that r (and thus x) depends on which dm we pick. Although m is our integration variable, we know the limits of the bar in terms of x, which can take values between $-L/2$ and $L/2$. To exploit the fact that we know the limits of x, we need to express dm in terms of x (i.e., we need to change the variable of integration.) To achieve this, we again use the object's density, which in the case of the rod will be the linear density λ, defined as mass per length. Then for each dm, we will have:

$$\lambda = \frac{dm}{dx} \Rightarrow dm = \lambda dx. \tag{4.16}$$

Since we have effectively expressed dm in terms of x, we can now rewrite the moment of inertia integral in terms of the single variable x:

$$I_O = \int r^2 dm = \int_{-L/2}^{L/2} x^2 \lambda dx. \tag{4.17}$$

Since the rod is uniform, λ is constant and it can come out of the integral. We then have:

$$I_O = \lambda \int_{-L/2}^{L/2} x^2 dx = \frac{\lambda}{3} \left(\frac{L^3}{8} + \frac{L^3}{8} \right) = \frac{1}{12} \lambda L^3. \tag{4.18}$$

However, because the rod is uniform of mass M and length L, we can also say that $\lambda = M/L$. Substituting this expression for λ into Equation (4.18) gives:

$$I_O = \frac{1}{12} \frac{M}{L} L^3 = \frac{1}{12} ML^2. \tag{4.19}$$

(b) If our axis of rotation goes through end A, the axis of rotation is not the same as the y axis, which we have defined to go through the midpoint. Of course, we could now redefine our y axis so that it again coincides with the axis of rotation. In that case, x would be between 0 and L instead of between -L/2 and L/2 and we could do our integral just like in part (a). However, let's just assume that we do not want to redefine our y axis. How can we set up the integral?

In this case, we see that if we arbitrarily pick two infinitesimal masses dm_1 and dm_2, $r_1 \neq x_1$ and $r_2 \neq x_2$, as shown in Figure 4.6b. In fact, what we see is that $r = (\frac{L}{2} + x)$. This expression works when $x > 0$ for the right half of the rod and when $x < 0$, for the left half of the rod. We can substitute this expression for r into our integral for the moment of inertia (and that $dm = \lambda dx$):

$$I_A = \int r^2 dm = \int_{-L/2}^{L/2} \left(\frac{L}{2} + x \right)^2 \lambda dx. \tag{4.20}$$

We have once again changed our variable of integration to be x and can easily see that $-L/2 \leq x \leq L/2$. Because λ is constant it can come out of the integral. We then have:

$$I_A = \lambda \int_{-L/2}^{L/2} \left(\frac{L}{2} + x \right)^2 dx$$

$$= \lambda \int_{-L/2}^{L/2} \left(\frac{L^2}{4} + Lx + x^2 \right) dx$$

$$= \lambda \left(\frac{L^2}{4} x + L \frac{x^2}{2} + \frac{x^3}{3} \right) \Big|_{-L/2}^{L/2} = \lambda \frac{L^3}{3}. \tag{4.21}$$

However, as we discussed earlier, $\lambda = M/L$. By substituting this expression to our result we get that:

$$I_A = \frac{1}{3}ML^2. \tag{4.22}$$

Let's think about why the moments of inertia about these two axes are different. Why is $I_A > I_O$? We reason in the same way as for the ring versus disk scenario. We see that in the case where the axis of rotation passes through the center of the rod, the infinitesimal masses dm of the rod will be located at distances between 0 and $L/2$ from the axis of rotation ($0 \le r \le L/2$). Notice that because in the definition of the moment of inertia r is the *distance* from the axis of rotation and it is squared, it does not matter if a mass is located to the left or right of the axis of rotation. However, when the axis of rotation is at one end of the rod, half of the rod is at a distance $L/2 \le r \le L$ from the axis of rotation. Therefore, because we have more mass further from the axis of rotation, the moment of inertia in this case will be greater. Indeed, it is much harder to get a rod to rotate about an axis that goes through one of its ends than to get it to rotate about an axis that goes through its center. To get the same angular acceleration, we will have to provide more net torque when the axis goes through one of the ends of the rod.

4.2 Some More Cool Theorems

There are some interesting theorems that can prove useful when finding moments of inertia, but as with the set of theorems we introduced in Chapter 2, these theorems are not needed to solve problems. We present these theorems at the end of this chapter for the sake of completeness of this unit on the moment of inertia, and because it is beneficial to know that they exist in the literature.

4.2.1 Theorem 1: Parallel Axis Theorem (Steiner Theorem)

The idea behind this theorem is that if the moment of inertia of a rigid body about an axis of rotation z that goes through its center of mass (I_{cm}) is known, then the moment of inertia about any other *parallel* axis z' that is at a distance d from z can be found via the relationship:

$$I_{z'} = I_{cm} + Md^2, \tag{4.23}$$

where M is the mass of the rigid body. There are two important notes associated with this theorem: (a) This relationship is true even if the axis z' is located outside the rigid body and (b) this relationship implies that the minimum moment of inertia of a rigid body is that about an axis of rotation that is going through its center of mass, since $Md^2 > 0$. But what is the proof of this relationship?

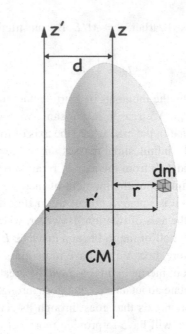

Figure 4.7 If the moment of inertia about an axis that goes through the rigid body's center of mass is known, then the moment of inertia about another parallel axis located at a distance d can be found easily using the parallel axis theorem.

Proof: Let's assume that in Figure 4.7 we are looking at one particular infinitesimal mass element dm of the rigid body. This dm has a moment of inertia about z, the axis that goes through the center of mass, equal to $r^2 dm$ where r is, as always, the perpendicular distance from dm to the z axis. Then the overall moment of inertia of the rigid body about the z axis, based on Equation (4.4), is $I_{cm} = \int r^2 dm$, where the integral is over all dm elements which make up the entire body.

Similarly, if we know that r' is the perpendicular distance between dm and axis z', then the moment of inertia of the rigid body about axis z' is $I_{z'} = \int r'^2 dm$. But what is the relationship between r and r'? From Figure 4.7 we see that $r' = r + d$. By substituting this result into our expression for $I_{z'}$, we get:

$$I_{z'} = \int r'^2 dm = \int (r+d)^2\, dm = \int \left(r^2 + 2rd + d^2 \right) dm. \qquad (4.24)$$

We can now split the integral into three separate integrals:

$$I_{z'} = \int r^2 dm + \int 2rd\, dm + \int d^2 dm. \qquad (4.25)$$

The first integral on the right-hand side is simply I_{cm}. Since the distance d between the axes z and z' is fixed, it can come out of the second and third integrals. Then we have:

$$I_{z'} = I_{cm} + 2d \int r\,dm + d^2 \int dm. \tag{4.26}$$

Let's think about these last two integrals. Firstly, we recall the definition of the center of mass of an object, which tells us the point about which the mass-weighted position is zero. Based on this definition, and since we are integrating over the dm elements with respect to an axis *through the center of mass*, we can see that $\int r\,dm = 0$. As for the second integral, we simply have that $\int dm = M$. Therefore, our final result is:

$$I_{z'} = I_{cm} + Md^2. \tag{4.27}$$

In Example 4 where we found the moment of inertia of the uniform rod about two different axes, we could use the result of part (a) and the parallel axis theorem to get the answer to part (b) without having to do any integration. We thought about making this an exercise at the end of the chapter, but it would be too easy!

4.2.2 Theorem 2: Perpendicular Axis Theorem

There are two versions of the perpendicular axis theorem. One involves planar objects and is the one most frequently discussed in textbooks. The other version is the generalization to three-dimensional (3D) objects. We will start with the simpler case of planar objects.

The Perpendicular Axis Theorem for Planar Objects

Let's look at a planar object such as the rectangular lamina in Figure 4.8. Imagine an axis z passing through the object and perpendicular to its plane. The moment of inertia I_z is the sum of the moments of inertia about two axes, x

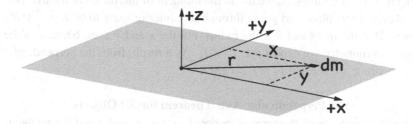

Figure 4.8 The moment of inertia about the z axis for this planar object is the sum of the moments of inertia about the x and y axes. Note that all axes go through the same point.

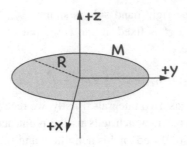

Figure 4.9 The moment of inertia about the z axis for this disk is the sum of the moments of inertia about the x and y axes. Because of symmetry, $I_x = I_y$. Therefore, $I_z = I_x + I_y = 2I_x = 2I_y$.

and y, both perpendicular to this z axis. All three axes pass through the same point, but the x and y axes are in the same plane as the object. Expressed as an equation,

$$I_z = I_x + I_y. \tag{4.28}$$

Proof: In Figure 4.8 we identify an infinitesimal mass element dm of the lamina. Its moments of inertia about the x, y, and z axes are $y^2 dm$, $x^2 dm$, and $r^2 dm$, respectively. If we now look at the overall moment of inertia about each axis, based on Equation (4.4) we have that $I_x = \int y^2 dm$, $I_y = \int x^2 dm$, and $I_z = \int r^2 dm$, where the integrals are over all dm elements that make up the lamina. We also see that $r^2 = x^2 + y^2$, and can substitute this result into our expression for I_z:

$$I_z = \int r^2 dm = \int \left(x^2 + y^2 \right) dm = \underbrace{\int x^2 dm}_{I_y} + \underbrace{\int y^2 dm}_{I_x} = I_x + I_y. \tag{4.29}$$

As an example of how this can be applied, let's again look at a uniform disk (Figure 4.9). In Example 3, we found the moment of inertia about the axis perpendicular to its plane and going through its center of mass to be $I_z = \frac{1}{2}MR^2$, where M is the mass and R is the radius. For the x and y axes, because of the object's symmetry, we expect that $I_x = I_y$. As a result, from the perpendicular axis theorem, we see that $I_x = I_y = \frac{1}{4}MR^2$.

The Perpendicular Axis Theorem for 3D Objects

For a 3D object, given the three orthogonal axes x, y, and z and the moments of inertia of the object about the three axes I_x, I_y, and I_z, respectively, the most general form of the perpendicular axis theorem is:

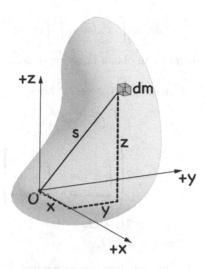

Figure 4.10 For a 3D rigid body, we look at an infinitesimally small mass element dm which is a distance s from the origin of the Cartesian coordinate system, with $s^2 = x^2 + y^2 + z^2$.

$$I_x + I_y + I_z = 2 \int s^2 dm. \tag{4.30}$$

As shown in Figure 4.10, s is the distance of the infinitesimal mass element dm of the 3D rigid object from the origin of the coordinate system, with $s^2 = x^2 + y^2 + z^2$.

Proof: Let's focus on the moment of inertia about the x axis (I_x) of the object in Figure 4.10. To find I_x for dm, we need to know the *perpendicular* distance of dm from the x axis. This distance will be $y^2 + z^2$. By accounting for all the dm elements that make up the 3D object, we have $I_x = \int \left(y^2 + z^2 \right) dm$. We can think in a similar way for the moments of inertia about the y and z axis and get the expressions $I_y = \int \left(x^2 + z^2 \right) dm$ and $I_z = \int \left(x^2 + y^2 \right) dm$, respectively. By adding together these three expressions we get:

$$I_x + I_y + I_z = \int \left(y^2 + z^2 \right) dm + \int \left(x^2 + z^2 \right) dm + \int \left(x^2 + y^2 \right) dm. \tag{4.31}$$

Now we can combine the three integrals and complete the proof:

$$I_x + I_y + I_z = \int \left(2x^2 + 2y^2 + 2z^2 \right) dm = 2 \int \underbrace{\left(x^2 + y^2 + z^2 \right)}_{s^2} dm = 2 \int s^2 dm. \tag{4.32}$$

From this derivation, we can now see how Equation (4.28) is the special case for a planar object. If we have an object in the xy plane, then $z = 0$ for every dm. Equation (4.31) then becomes:

$$I_x + I_y + I_z = \int y^2 dm + \int x^2 dm + \int \left(x^2 + y^2\right) dm. \qquad (4.33)$$

We can combine the integrals and obtain Equation (4.28):

$$\begin{aligned}
I_x + I_y + I_z &= \int \left(2x^2 + 2y^2\right) dm \\
&= 2 \int \left(x^2 + y^2\right) dm \\
&= 2 \left(\underbrace{\int x^2 dm}_{I_y} + \underbrace{\int y^2 dm}_{I_x} \right) \\
&= 2I_x + 2I_y \\
\Rightarrow I_z &= I_x + I_y. \qquad (4.34)
\end{aligned}$$

As an example of how this equation can be applied, let's look at a hollow sphere (i.e., a spherical shell) of mass M and radius R. If we make the center of the shell the origin of the coordinate system, because of symmetry we expect that $I_x = I_y = I_z$. Then, Equation (4.30) gives:

$$3I_x = 3I_y = 3I_z = 2 \int s^2 dm. \qquad (4.35)$$

However, each dm that makes up the shell is the same distance R away from the origin (i.e., $s = R$). Then we have:

$$3I_x = 3I_y = 3I_z = 2 \int R^2 dm = 2R^2 \underbrace{\int dm}_{M} = 2MR^2$$

$$\Rightarrow I_x = I_y = I_z = \frac{2}{3}MR^2. \qquad (4.36)$$

In general, we see that these theorems provide an alternative way of finding the moments of inertia about different axes, especially if we can exploit an object's symmetry.

4.3 Moment of Inertia of Composite Areas

A composite area can be made by combining simple-shaped areas like squares, circles, and rectangles (e.g., a circle inscribed in a square). In many cases, we already know the moment of inertia of each simple shape about an axis of rotation through its own center of mass, as these results are commonly tabulated. But what if we want to find the moment of inertia of the composite shape about some arbitrary axis of rotation? Rather than doing another (often complicated) integral, we can use the parallel and/or perpendicular axis theorem(s). Since

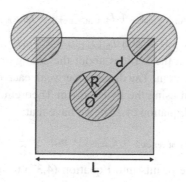

Figure 4.11 A composite area consisting of a uniform thin square of side length
L and three uniform thin disks of radius R. One disk is placed at the center of the
square O, and the other two are placed such that their centers coincide with two
consecutive vertices.

the moments of inertia of the constituent areas are known about their own centers of mass, it is straightforward to use one or both of these theorems to find
the moment of inertia of each shape about the new axis of rotation. Then, we
can algebraically sum all these moments to find the moment of inertia of the
composite shape. If there are any holes in the area, then the moment of inertia
associated with the missing mass should be subtracted, since it does not contribute to the moment of inertia of the composite object. Let's do an example to
illustrate this idea.

Example 5. The moment of inertia of a composite object. A planar object
consists of a uniform thin square of side length L and mass M and three uniform thin disks, each of radius R and mass m. The disks are placed so that
one is on top of the square, with its center O coinciding with the center of
the square. The other two disks are placed so that their centers coincide with
two consecutive vertices of the square. The configuration is shown in Figure
4.11. Find the moment of inertia of this composite object about an axis that
goes through the center of the square O and is perpendicular to the plane of the
object (and thus perpendicular to the plane of the page). The moment of inertia
of a square and a disk about an axis that goes through each object's center of
mass and is perpendicular to its plane is $I_{\text{cm.square}} = \frac{1}{6}ML^2$ and $I_{\text{cm.disk}} = \frac{1}{2}mR^2$,
respectively.

   In this example, we have a composite area that consists of one square and
three disks. To find the moment of inertia about an axis that goes through O and
is perpendicular to the plane of the page, we simply have to add each object's
moment of inertia about the same axis:

$$I_O = I_{\text{square}} + I_{\text{disk at center}} + I_{\text{disk at vertex 1}} + I_{\text{disk at vertex 2}}. \qquad (4.37)$$

We know that $I_{\text{square}} = \frac{1}{6}ML^2$ and $I_{\text{disk at center}} = \frac{1}{2}mR^2$. To find the moments of inertia of the disks at the vertices about the axis going through O we can use the parallel axis theorem. The distance between each disk's center and O is $d = \frac{L}{\sqrt{2}}$, which we find using the Pythagorean Theorem. Therefore, using the parallel axis theorem (Equation (4.23)), we have that:

$$I_{\text{disk at vertex 1}} = I_{\text{disk at vertex 2}} = I_{\text{cm.disk}} + md^2 = \frac{1}{2}mR^2 + m\frac{L^2}{2}. \qquad (4.38)$$

We can now plug these results into Equation (4.37) to obtain the moment of inertia of this composite object:

$$I_O = \frac{1}{6}ML^2 + \frac{1}{2}mR^2 + \left(\frac{1}{2}mR^2 + m\frac{L^2}{2}\right) + \left(\frac{1}{2}mR^2 + m\frac{L^2}{2}\right)$$

$$= \frac{1}{6}ML^2 + \frac{3}{2}mR^2 + mL^2. \qquad (4.39)$$

Exercises

(i) Compute the moment of inertia of (a) a uniform cylindrical shell and (b) a uniform solid cylinder, both with length L, radius R, and mass M, about an axis passing through their center along their length. How do your answers compare to what we found for the ring and uniform thin disk?

(ii) Let's take another look at the uniform rod in Example 4. In both parts of that example, we were concerned with relating r from the definition of the moment of inertia to x of our coordinate system. For this exercise, compute the moment of inertia of the thin rod about the same two axes (one which passes through the rod's midpoint and another which passes through one of the ends of the rod) by integrating in terms of r rather than x.

(iii) A thin rod 10 m long has a density that varies linearly from 2 to 22 kg/m. Find (a) its mass M, (b) the position of its center of mass x_{CM}, (c) the moment of inertia I about an axis perpendicular to the rod that passes through x_{CM}, and (d) the moment of inertia I about an axis perpendicular to the rod that passes through the heavy end.

(iv) Compute the moment of inertia of the thin disk in Figure 4.9 about either the x or y axis by direct integration (i.e., using $I = \int r^2 dm$). Verify that this result is the same as what we obtain using symmetry arguments and the perpendicular axis theorem.

(v) Let's look again at the uniform, rectangular plate shown in Figure 4.8. Assume it has mass M, length a along the x axis, and length b along the y axis. The coordinate axes defined in this figure are called the principal axes of rotation and the corresponding moments of inertia are called principal moments of inertia (we will briefly discuss this idea in Chapter 8). Show that the principal moments of inertia about the x, y, and z axes are $I_x = \frac{M}{12}b^2$, $I_y = \frac{M}{12}a^2$, and $I_z = \frac{M}{12}(a^2 + b^2)$, respectively. Find the moment of inertia about the z axis in two different ways – one by direct integration and the other by using the perpendicular axis theorem (assuming you first compute the moment of inertia about the x and y axes).

(vi) A thin square of side length $2l$ has a cutout in the shape of a circle of radius l (i.e., the circle is inscribed in the square). Compute the moment of inertia of this object about an axis of rotation intersecting the center of the circular cutout and perpendicular to its plane.

5
Angular Momentum

Our discussion so far has focused on Newton's Laws of motion and their extension to rotational motion. However, in our first course in mechanics when studying linear motion, we also learned about momentum, work, and energy. But why did we need to learn about these concepts? Aren't Newton's Laws enough for studying all mechanics?

While it is true that Newton's laws describe the resulting motion when we know the forces that are being exerted on a system, the concepts of momentum, work, and energy (a) allow us to study phenomena where forces are difficult to find and quantify (e.g., in collisions) and (b) due to symmetries we see in nature, lead to conservation laws that are very powerful and useful to have in our toolkit. Therefore, we will follow the same trajectory here and define the rotational analogs of these quantities, starting with the concept of angular momentum.

5.1 Review: Linear Momentum ($\vec{p}$)

Let's recall that the linear momentum ($\vec{p}$) is a property of all moving masses and is defined as the product of an object's mass (m) and velocity ($\vec{v}$): $\vec{p} = m\vec{v}$, with the SI unit of [kgm/s]. We can use the change in an object's momentum to determine the net force ($\Sigma\vec{F}$) exerted on it. Alternatively, an object's change in momentum can be found if we know the net force. This is famously described in the general form of Newton's Second Law, which quantifies the relationship between the net force on an object and the rate of change of its momentum:

$$\Sigma\vec{F} = \frac{d\vec{p}}{dt}. \tag{5.1}$$

From Equation (5.1) we see that if the net force is constant, then the rate of change of the momentum is constant. If there is no net force, then the momentum does not change (so its rate of change is zero). The momentum of an object changes only when a net force is exerted on it by another object.

For example, if we throw a ball toward a window, the window will exert a force on the ball when it hits the window, and so the ball's momentum will change. Because of Newton's Third Law, there is a force equal in magnitude but opposite in direction exerted by the ball on the window. If we want to predict how much force the ball will exert on the window, and thus whether or not the window will break, we can perform a calculation to predict the ball's change in momentum rather than actually throw the ball at the window.

It is possible to speak in a similar way about bodies in rotational motion and the effect they have on objects with which they come into contact. For example, imagine we want to quantify the change in a tennis ball's spin as it is hit by a racket. The quantity that describes this effect for rotational motion is called angular momentum, $\vec{L}$. But how do we define angular momentum?

5.2 Definition of Angular Momentum ($\vec{L}$)

The linear momentum is defined in the most general expression of Newton's Second Law as the quantity that changes with time as a result of a net force $\left(\Sigma \vec{F} = \frac{d\vec{p}}{dt} \right)$. Let's assume the angular momentum $\vec{L}$ is the rotational equivalent – that is, it is the quantity that changes with time as a result of a net torque. Thus, we seek the quantity $\vec{L}$ such that the most general expression of Newton's Second Law for rotational motion can be written as:

$$\Sigma \vec{\tau} = \frac{d\vec{L}}{dt}. \qquad (5.2)$$

Let's start by looking at a point mass m moving with respect to an axis z, as shown in Figure 5.1. The trajectory of the mass m does not have to be circular for us to say that the mass is rotating about axis z. This is because, relative to axis z, the mass has an angular position that changes with time. This is true even if m is moving linearly, which we can illustrate as follows: Imagine sitting on the side of a straight road and watching a car pass by you. For you to keep your eyes on the car as it goes down the road, you will have to rotate your head. To you, it appears as though the car is rotating around you, and so your body is the axis of rotation. With this in mind, we can now see that Figure 5.1 can represent any motion of the mass in the xy plane.

Let's say that at some time t we find mass m to have a position vector $\vec{r}$ with respect to the z axis, a linear velocity $\vec{v}$, and a linear momentum $\vec{p}$, as shown in Figure 5.1. As always, $\vec{r}$ is perpendicular to the axis of rotation z. We know that if $\vec{p}$ changes in magnitude or direction, then from Newton's Second Law there is a net force $\Sigma \vec{F}$ whose direction is the same as that of the change in momentum. This net force will produce a net torque about axis z:

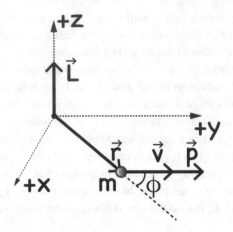

Figure 5.1 A mass m is moving in the xy plane with position vector $\vec{r}$ relative to the z axis. At any time t, the mass will have a linear momentum $\vec{p}$ and an angular momentum $\vec{L}$ about axis z, given by the right-hand rule.

$$\Sigma \vec{\tau} = \vec{r} \times \Sigma \vec{F} = \vec{r} \times \frac{d\vec{p}}{dt}, \qquad (5.3)$$

where we have substituted in Equation (5.1) for $\Sigma \vec{F}$. Now let's make some clever substitutions. Keep in mind that, by the definition of linear momentum, $\vec{p}$ has the same direction as $\vec{v} = \frac{d\vec{r}}{dt}$. This means that:

$$\vec{v} \times \vec{p} = \frac{d\vec{r}}{dt} \times \vec{p} = 0, \qquad (5.4)$$

because the cross product of two vectors that are parallel to each other is zero. Adding $\frac{d\vec{r}}{dt} \times \vec{p} = 0$ to the right hand side of Equation (5.3) gives:

$$\Sigma \vec{\tau} = \vec{r} \times \frac{d\vec{p}}{dt} + 0$$

$$= \vec{r} \times \frac{d\vec{p}}{dt} + \frac{d\vec{r}}{dt} \times \vec{p}$$

$$= \frac{d(\vec{r} \times \vec{p})}{dt}. \qquad (5.5)$$

By comparing Equation (5.5) to Equation (5.2), we see that angular momentum is defined as[1]:

$$\vec{L} = \vec{r} \times \vec{p}, \qquad (5.6)$$

---

[1] In the most general case $\vec{L} = \vec{r} \times \vec{p}$ is defined relative to a *point*, not an axis. Since we are studying rotation about fixed axes in this book, we are looking at the specific case of defining $\vec{L}$ with respect to an axis of rotation, and thus we always use the component of $\vec{r}$ perpendicular to the fixed axis. We will briefly look at the definition of angular momentum with respect to a point in Chapter 8.

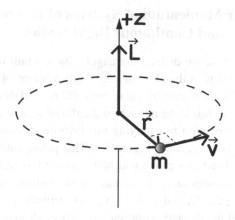

Figure 5.2 A point mass m is executing circular motion on the xy plane about the z axis that goes through the center of the circle. In this case, the linear velocity $\vec{v}$ is always perpendicular to the position vector $\vec{r}$. Applying the right-hand rule shows that $\vec{L}$ points along the $+z$ axis.

with SI unit of [kgm^2/s]. Thus, the most general form of Newton's Second Law for rotational motion can be written as $\Sigma\vec{\tau} = \frac{d\vec{L}}{dt}$ where $\vec{L} = \vec{r} \times \vec{p}$.

For the point mass, $\vec{p} = m\vec{v}$, so Equation (5.6) gives the magnitude of the angular momentum:

$$|\vec{L}| = |\vec{r} \times \vec{p}| = rp \sin\phi = rmv \sin\phi. \tag{5.7}$$

If $\vec{r}$ is parallel or anti-parallel to $\vec{p}$, then $\phi = 0$ or $\phi = \pi$, respectively, and $\vec{L} = 0$. Since $\vec{L}$ is the result of a cross product, to get its direction we use the right hand rule where we point the fingers of our right hand in the direction of $\vec{r}$, bend our fingers to align them with the direction $\vec{p}$, and our extended thumb then gives the direction of $\vec{L}$. For the situation shown in Figure 5.1, $\vec{L}$ will be along the $+z$ axis as shown.

So far we have looked at a point mass executing an arbitrary trajectory. What if the trajectory is circular? In this case, as shown in Figure 5.2, $\vec{v}$ is perpendicular to $\vec{r}$ and $\phi = \frac{\pi}{2}$. Therefore, Equation (5.7) becomes:

$$|\vec{L}| = rmv \sin\phi = rmv \sin\left(\frac{\pi}{2}\right) = rmv. \tag{5.8}$$

Application of the right-hand rule to the scenario in Figure 5.2 gives that $\vec{L}$ points upwards along the $+z$ axis.

5.3 Angular Momentum of Systems of Discrete Masses and Continuous Rigid Bodies

In the previous section, we defined the angular momentum for a point mass. Our goal is now to extend this definition to study systems of discrete masses and rigid bodies. The derivation in this section will remind us of our derivation in Chapter 4 when we found the moment of inertia of such systems.

Let's start the derivation by studying our favorite arbitrarily shaped rigid body, the potato, shown again in Figure 5.3. This potato rotates with angular velocity $\vec{\omega}$ about a fixed axis z. We will first consider this rigid body as a collection of n small masses m_1, m_2, ..., m_n, whose position vectors and linear velocities are $\vec{r}_1$, $\vec{r}_2$,..., $\vec{r}_n$, and $\vec{v}_1$, $\vec{v}_2$,...,$\vec{v}_n$, respectively. Because the potato is a rigid body and the distance between any two small masses remains constant, it must be the case that every small mass rotates with the same angular velocity $\vec{\omega}$.

Because all small masses rotate about the same axis of rotation, for the object as a whole, we will find the total angular momentum ($\vec{L}_{\text{tot}}$) by taking the vector sum of the angular momenta of the masses:

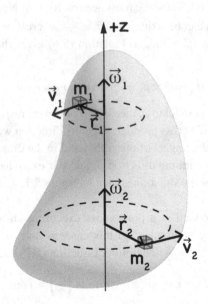

Figure 5.3 A potato of mass M represents any arbitrarily shaped rigid object. This object is rotating about an axis z that remains fixed. The potato consists of many small masses m_1, m_2, etc. whose position vectors and linear velocities are $\vec{r}_1$, $\vec{r}_2$, etc., and $\vec{v}_1$, $\vec{v}_2$, etc., respectively. Since the potato is rigid, all small masses have the same angular velocity $\vec{\omega}_1 = \vec{\omega}_2 = ... = \vec{\omega}$. The potato's angular momentum can be shown to be $\vec{L} = I\vec{\omega}$, where I is its moment of inertia.

$$\vec{L}_{tot} = \vec{L}_1 + \vec{L}_2 + ... + \vec{L}_n$$
$$= \vec{r}_1 \times \vec{p}_1 + \vec{r}_2 \times \vec{p}_2 + ... + \vec{r}_n \times \vec{p}_n, \tag{5.9}$$

where we have substituted Equation (5.6) for the angular momentum of each small mass. Using $\vec{p} = m\vec{v}$ and $\vec{v} = \vec{\omega} \times \vec{r}$, Equation (5.9) becomes:

$$\vec{L}_{tot} = \vec{r}_1 \times (m_1\vec{v}_1) + \vec{r}_2 \times (m_2\vec{v}_2) + ... + \vec{r}_n \times (m_n\vec{v}_n)$$
$$= m_1\vec{r}_1 \times \vec{v}_1 + m_2\vec{r}_2 \times \vec{v}_2 + ... + m_n\vec{r}_n \times \vec{v}_n$$
$$= m_1\vec{r}_1 \times (\vec{\omega}_1 \times \vec{r}_1) + m_2\vec{r}_2 \times (\vec{\omega}_2 \times \vec{r}_2) + ... + m_n\vec{r}_n \times (\vec{\omega}_n \times \vec{r}_n). \tag{5.10}$$

We see that each term in the last expression contains a triple cross product. To continue our derivation, we use the identity $\vec{A} \times (\vec{B} \times \vec{C}) = \vec{B}(\vec{A} \cdot \vec{C}) - \vec{C}(\vec{A} \cdot \vec{B})$. We then have for mass i:

$$\vec{r}_i \times (\vec{\omega}_i \times \vec{r}_i) = \vec{\omega}_i \underbrace{(\vec{r}_i \cdot \vec{r}_i)}_{r_i^2} - \vec{r}_i \underbrace{(\vec{r}_i \cdot \vec{\omega}_i)}_{0} = r_i^2 \vec{\omega}_i, \tag{5.11}$$

where $\vec{r}_i \cdot \vec{\omega}_i = 0$ because $\vec{\omega}_i$ is always perpendicular to $\vec{r}_i$. By substituting the result of Equation (5.11) into Equation (5.10) we obtain an expression for the total angular momentum of a system of discrete small masses:

$$\vec{L}_{tot} = m_1 r_1^2 \vec{\omega}_1 + m_2 r_2^2 \vec{\omega}_2 + ... + m_n r_n^2 \vec{\omega}_n. \tag{5.12}$$

In the case of the rigid body, we note that $\vec{\omega}_1 = \vec{\omega}_2 = ... = \vec{\omega}_n = \vec{\omega}$. Therefore, the result of Equation (5.12) can be further simplified:

$$\vec{L}_{tot} = m_1 r_1^2 \vec{\omega} + m_2 r_2^2 \vec{\omega} + ... + m_n r_n^2 \vec{\omega}$$
$$= (m_1 r_1^2 + m_2 r_2^2 + ... + m_n r_n^2) \vec{\omega}$$
$$= \left(\Sigma_{i=1}^n m_i r_i^2 \right) \vec{\omega}. \tag{5.13}$$

As we know, the last sum is the moment of inertia of a system of discrete small masses. However, the potato is really a continuous object. Therefore, we can make this sum more accurate by considering the limit where the masses become infinitesimal: $dm_1, dm_2, ..., dm_n$, so that the sum becomes an integral over m, and we are left with the angular momentum of a rigid body:

$$\vec{L} = \left(\int r^2 dm \right) \vec{\omega} = I\vec{\omega}. \tag{5.14}$$

Equations (5.13) and (5.14) can also be applied to find the angular momentum of a point mass with $I = mr^2$. In this case, Equation (5.13) reduces to $\vec{L} = mr^2 \vec{\omega} = I\vec{\omega}$.

Equation (5.14) shows that for rotation about a fixed axis, since I is a positive scalar and $\vec{\omega}$ is a vector, the angular momentum is also a vector whose magnitude is $L = I\omega$ and whose direction is that of $\vec{\omega}$. This means that since $\vec{\omega}$ is perpendicular to the plane of rotation and along the axis of rotation, so is the angular momentum $\vec{L}$. As we discussed in Chapter 1, the convention for the direction of $\vec{\omega}$, and therefore $\vec{L}$, allows us to fully describe rotational motion via vectors that are normal to the plane of rotation and, thus, to also reduce a two-dimensional motion to one dimension.

The result of Equation (5.14) is somewhat expected. We could have easily found $\vec{L}$ by using the equation for $\vec{p} = m\vec{v}$ and substituting the rotational analogs I and $\vec{\omega}$ for m and $\vec{v}$, respectively. Although obtaining $\vec{L}$ in this manner would have been consistent with what we have said thus far about drawing analogies from our knowledge of translational motion, it would not have been a mathematically satisfying derivation.

5.4 Change in Angular Momentum due to External Torques

Let's return to an idea we developed when studying linear motion: Recall that we can classify forces on objects in a system based on whether or not the source of the force is within the system (internal force) or outside the system (external force). The net force on the system can then be written as the sum of all the internal and external forces:

$$\Sigma\vec{F} = \vec{F}_{\text{int.1}} + \vec{F}_{\text{int.2}} + \ldots + \vec{F}_{\text{ext.1}} + \vec{F}_{\text{ext.2}} + \ldots = \Sigma\vec{F}_{\text{int}} + \Sigma\vec{F}_{\text{ext}}. \quad (5.15)$$

However, $\Sigma\vec{F}_{\text{int}} = 0$ because of Newton's Third Law. Taking the cross product of $\vec{r} \times \Sigma\vec{F}$ using Equation (5.15) we get:

$$\vec{r} \times \Sigma\vec{F} = \vec{r} \times \underbrace{\Sigma\vec{F}_{\text{int}}}_{0} + \vec{r} \times \Sigma\vec{F}_{\text{ext}}$$

$$\Rightarrow \Sigma\vec{\tau} = \Sigma\vec{\tau}_{\text{ext}}. \quad (5.16)$$

So, for a system of point masses or a rigid body, Newton's Second Law for rotational motion becomes:

$$\Sigma\vec{\tau} = \Sigma\vec{\tau}_{\text{ext}} = \frac{d\vec{L}_{\text{tot}}}{dt}. \quad (5.17)$$

This result implies that only external torques change the system's angular momentum $\vec{L}_{\text{tot}}$. The torques due to internal forces cancel themselves out for a system of objects because the internal forces cancel themselves out. It will be useful for us to remember this equation when we derive the Conservation of Angular Momentum theorem later in this chapter.

5.5 Newton's Second Law for Rotational Motion with Constant Moment of Inertia

Recall that in the case of translational motion, $\Sigma \vec{F} = \frac{d\vec{p}}{dt}$ becomes $\Sigma \vec{F} = m\vec{a}$ when m is constant. Since the equivalent of m for rotational motion is I, can we simplify the expression $\Sigma \vec{\tau} = \frac{d\vec{L}}{dt}$ for constant I?

The answer is yes, and the derivation is very similar to the one we did for translational motion. We start with the general expression of Newton's Second Law for rotational motion and substitute in Equation (5.14), which expresses the angular momentum in terms of the moment of inertia:

$$\Sigma \vec{\tau} = \frac{d\vec{L}}{dt} = \frac{d(I\vec{\omega})}{dt} = \frac{dI}{dt}\vec{\omega} + I\frac{d\vec{\omega}}{dt}. \tag{5.18}$$

If I is constant, then $\frac{dI}{dt} = 0$. So the first term disappears, and using $\frac{d\vec{\omega}}{dt} = \vec{\alpha}$ we are left with

$$\Sigma \vec{\tau} = I\vec{\alpha}, \tag{5.19}$$

which we studied extensively in Chapter 3. It is important to repeat that this expression is true only if I is constant, just like $\Sigma \vec{F} = m\vec{a}$ is true only if m is constant. As we discussed in Chapter 3, given the equivalencies between force and torque, mass and moment of inertia, and linear and angular acceleration, the equivalent of $\Sigma \vec{F} = m\vec{a}$ is $\Sigma \vec{\tau} = I\vec{\alpha}$.

5.6 Notes on Angular Momentum and Newton's Second Law

Before we present some examples, let's open a parenthesis to discuss some subtleties associated with angular momentum and Newton's Second Law for rotational motion.

(i) As we discussed in Chapter 1 for the case of velocity and acceleration, if we have a quantity A whose rate of change is constant (i.e., $\frac{dA}{dt}$ = constant), then the average rate of change $\frac{\Delta A}{\Delta t}$ equals the instantaneous rate of change $\frac{dA}{dt}$. That is $\frac{dA}{dt} = \frac{\Delta A}{\Delta t}$. Therefore, if $\Sigma \vec{\tau} = \frac{d\vec{L}}{dt}$ is constant, meaning that the net torque is constant, then

$$\Sigma \vec{\tau} = \frac{\Delta \vec{L}}{\Delta t} = \frac{\vec{L}_f - \vec{L}_i}{t_f - t_i}. \tag{5.20}$$

(ii) As mentioned, Equation (5.19) is true only if I is constant. If $I = \int r^2 dm$ changes, either because the mass of the object changes, the axis of rotation changes, or the distribution of the mass about the axis of rotation changes, then we cannot use this equation (other than to find

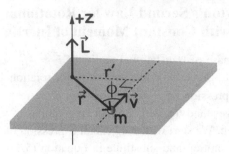

Figure 5.4 The angular momentum of mass m about the z axis is $\vec{L} = \vec{r} \times \vec{p} = m\vec{r} \times \vec{v}$. Its magnitude is $L = mrv \sin \phi = mvr'$, where $r' = r \sin \phi$ is the perpendicular distance from the axis z to m's line of motion.

instantaneous values). In these cases, one must use Equation (5.2) (or (5.20), for a constant net torque).

(iii) In Figure 5.4, the angular momentum of mass m about the z axis is $\vec{L} = \vec{r} \times \vec{p} \Rightarrow \vec{L} = m\vec{r} \times \vec{v}$. The magnitude is given by $L = mrv \sin \phi$ where r is the distance from the axis of rotation z to the location of the mass. However, a closer examination of Figure 5.4 shows that $r \sin \phi = r'$, where r' is the perpendicular distance between the axis z and the line of motion of the mass. Thus, $L = mvr'$. We saw something similar in Chapter 2 when using the moment arm to compute torque due to a force. This result will be helpful for one of the exercises at the end of the chapter.

Example 1: Change in angular momentum of a rotating disk. The uniform horizontal disk shown in Figure 5.5 is rotating about a vertical axis z that goes through its center O. At time t_1 the disk has angular velocity $\vec{\omega}_1$ of magnitude 10 rad/s and direction as shown. At time t_2, we find that the disk has angular velocity $\vec{\omega}_2$ of magnitude 30 rad/s in the opposite direction. The disk's moment of inertia about the z axis is $I_z = 0.5$ kgm^2. (a) Find the change in angular momentum. (b) If the change occurred at a constant rate during a time interval of $\Delta t = t_2 - t_1 = 5.0$ s, what is the net torque that caused this change in angular momentum?

(a) From Equation (5.14), we know that the angular momentum vector has the same direction as the angular velocity vector, as shown in Figure 5.5. We see that both angular momentum vectors are along the same axis z. Based on our coordinate system, $\vec{L}_1$ is positive while $\vec{L}_2$ is negative.

To find the change in angular momentum $\Delta \vec{L}$, since both vectors are along the z axis, we can work algebraically:

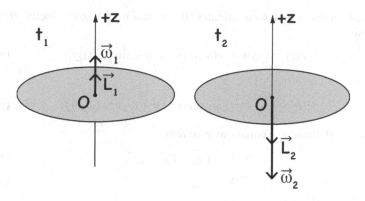

Figure 5.5 A uniform horizontal disk switches direction of rotation during a time interval of $\Delta t = t_2 - t_1 = 5.0$ s. The direction of the angular momentum changes because the direction of the angular velocity $\vec{\omega}$ changes. The change in angular momentum is $\Delta \vec{L} = \vec{L}_2 - \vec{L}_1$.

$$\Delta \vec{L} = \vec{L}_2 - \vec{L}_1$$
$$\Rightarrow \Delta L = -L_2 - (+L_1) = -I_2 \omega_2 - I_1 \omega_1. \tag{5.21}$$

However, $I_1 = I_2 = I$ because the axis of rotation has not changed and neither has the distribution of the disk's mass about this axis. Therefore, we now have:

$$\Delta L = -I(\omega_2 + \omega_1) = -0.5 \text{ kgm}^2(30 \text{ rad/s} + 10 \text{ rad/s}) = -20 \text{ kgm}^2/\text{s}. \tag{5.22}$$

Of course, the negative sign of our answer indicates that $\Delta \vec{L}$ is pointing along the $-z$ axis.

(b) Since the change in angular momentum occurred at a constant rate, we can use Equation (5.20). Because $\Delta t > 0$, we see that the change in angular momentum has the same direction as the net torque that caused it. Thus, $\Sigma \vec{\tau}$ is also pointing downwards, along the $-z$ axis. For the magnitude of the net torque we have:

$$\Sigma \tau = \frac{\Delta L}{\Delta t} = \frac{20}{5.0} = 4.0 \text{ Nm}. \tag{5.23}$$

Example 2: Angular momentum of a moving particle. A particle of mass m is moving along a trajectory in the xy plane and its coordinates at any time t are given by $(x, y) = (x_0 + \cos(\omega_1 t), y_0 + \sin(\omega_2 t))$, where x_0, y_0, ω_1, and ω_2 are constants. What is the angular momentum of the particle about the z axis at time $t = 0$?

Based on the x and y coordinates, the position $\vec{r}(t)$ and velocity $\vec{v}(t)$ are given by

$$\vec{r}(t) = (x_0 + \cos(\omega_1 t))\hat{i} + (y_0 + \sin(\omega_2 t))\hat{j}, \qquad (5.24)$$

and

$$\vec{v}(t) = \frac{d\vec{r}(t)}{dt} = -\omega_1 \sin(\omega_1 t)\hat{i} + \omega_2 \cos(\omega_2 t)\hat{j}. \qquad (5.25)$$

At time $t = 0$, the above equations give that

$$\vec{r}(0) = (x_0 + 1)\hat{i} + y_0\hat{j}, \qquad (5.26)$$

$$\vec{v}(0) = \omega_2\hat{j}. \qquad (5.27)$$

By definition, the angular momentum is thus given by

$$\begin{aligned}
\vec{L}(0) &= \vec{r}(0) \times \vec{p}(0) = \vec{r}(0) \times m\vec{v}(0) = m(\vec{r}(0) \times \vec{v}(0)) \\
&= m((x_0 + 1)\hat{i} + y_0\hat{j}) \times (\omega_2\hat{j}) \\
&= m((x_0 + 1)\omega_2)\underbrace{\hat{i} \times \hat{j}}_{\hat{k}} + my_0(\omega_2)\underbrace{\hat{j} \times \hat{j}}_{0} \\
&= m\omega_2(x_0 + 1)\hat{k}. \qquad (5.28)
\end{aligned}$$

Our final result confirms that the angular momentum is along the $+z$ axis, as would be expected since $\vec{r}(0)$ is in the xy plane and $\vec{v}(0)$ is along the y axis. Therefore, since $\vec{L}(0)$ is the cross product of the vectors $\vec{r}(0)$ and $\vec{v}(0)$, it has to be perpendicular to both these vectors. In fact, one can show that since both $\vec{r}(t)$ and $\vec{v}(t)$ are in the xy plane for all times, then $\vec{L}(t)$ will be along the z axis at all times. Showing this is part of an exercise at the end of the chapter.

5.7 Conservation of Angular Momentum

The theorem of Conservation of Momentum (COM) was one of the foundational principles that we first derived in our study of linear motion. Just as we used the most general form of Newton's Second Law to derive this theorem, it is reasonable to expect that we could use the most general form of Newton's Second Law for the rotational motion to derive an analogous principle for angular momentum.

5.7.1 The Momentum-Impulse Theorem and Conservation of Momentum for Translational Motion

Let's first remind ourselves of the derivation of the theorem of COM for linear motion. We start with the most general expression of Newton's Second Law to describe a system of objects. For a system of objects, due to Newton's Third Law, the internal forces cancel out, and any net force will be due to external forces only. Thus:

$$\Sigma \vec{F}_{\text{ext}} = \frac{d\vec{p}_{\text{tot}}}{dt}, \tag{5.29}$$

where $\vec{p}_{\text{tot}}$ is the vector sum of the linear momenta of all objects in the system. Now, let's multiply both sides with dt and take the integral of the equation. The variable of integration on the left side is t while on the right side is $\vec{p}_{\text{tot}}$. For the limits of integration, at time t_i we assume that the momentum is $\vec{p}_{\text{tot.i}}$ and at time t_f the momentum is $\vec{p}_{\text{tot.f}}$:

$$\int_{t_i}^{t_f} \Sigma \vec{F}_{\text{ext}} dt = \int_{\vec{p}_{\text{tot.i}}}^{\vec{p}_{\text{tot.f}}} d\vec{p}_{\text{tot}}. \tag{5.30}$$

At this point, we define the left-hand side of the equation to be equal to the impulse imparted by the net external force $\vec{\mathcal{I}}_{\text{ext}}$ during the time interval from t_i to t_f:

$$\vec{\mathcal{I}}_{\text{ext}} := \int_{t_i}^{t_f} \Sigma \vec{F}_{\text{ext}} dt. \tag{5.31}$$

Because $t_f > t_i$, it is clear from this definition that the direction of the impulse $\vec{\mathcal{I}}_{\text{ext}}$ will be the same as that of the net external force. Of course, the impulse $\vec{\mathcal{I}}$ (which is a vector) is not to be confused with the moment of inertia about an axis I (which is a scalar).

With this definition in mind, Equation (5.30) becomes:

$$\vec{\mathcal{I}}_{\text{ext}} = \int_{\vec{p}_{\text{tot.i}}}^{\vec{p}_{\text{tot.f}}} d\vec{p}_{\text{tot}} = \vec{p}_{\text{tot.f}} - \vec{p}_{\text{tot.i}} = \Delta\vec{p}_{\text{tot}}, \tag{5.32}$$

or

$$\vec{p}_{\text{tot.i}} + \vec{\mathcal{I}}_{\text{ext}} = \vec{p}_{\text{tot.f}}, \tag{5.33}$$

where we can see that $\vec{\mathcal{I}}_{\text{ext}}$ also has the same direction as the change in momentum $\Delta\vec{p}$.

Equation (5.33) is known as the Momentum-Impulse Theorem (MIT), which is really Newton's Second Law written in a different form. Since we started this derivation with the most general expression of Newton's Second Law, which is always true, MIT is also always true. It says that the total final linear momentum of a system of objects is equal to whatever the total initial momentum was plus any additional momentum (impulse) that was imparted to the system by the net external force. Since MIT is a vector equation, the final momentum $\vec{p}_{\text{tot.f}}$ will be the vector sum of $\vec{p}_{\text{tot.i}}$ and $\vec{\mathcal{I}}_{\text{ext}}$.

If the impulse $\vec{\mathcal{I}}_{\text{ext}}$ is zero (or small enough that we can approximate it to be zero), Equation (5.33) simplifies and we obtain the theorem of COM:

$$\vec{p}_{\text{tot.i}} = \vec{p}_{\text{tot.f}}, \tag{5.34}$$

which states that the total linear momentum remains constant in both magnitude and direction.

There are a couple of important points we need to make here. First, we repeat that while MIT is always true, COM is true only if the impulse due to external forces is zero or negligible. Second, as we went from the most general expression of Newton's Second Law (which is always true) to MIT, we did not require that the mass remain constant. Similarly, we did not require that the mass remain constant when we obtained COM from MIT. In other words, MIT and COM are useful to study systems in which the mass changes. On the other hand, $\Sigma \vec{F} = m\vec{a}$ is only applicable if m is constant.

5.7.2 The Angular Momentum-Impulse Theorem and Conservation of Angular Momentum

In this section, we will see that the derivation of the Conservation of Angular Momentum (COAM) parallels that of COM. We start with the most general expression of Newton's Second Law for rotational motion to describe a system of objects. Due to Newton's Third Law, the internal forces (and thus internal torques) again cancel out. Thus, any net torque will be due to external forces only:

$$\Sigma \vec{\tau}_{\text{ext}} = \frac{d\vec{L}_{\text{tot}}}{dt}, \tag{5.35}$$

where $\vec{L}_{\text{tot}}$ is the vector sum of the angular momenta of all objects in the system. Now, let's multiply both sides with dt and take the integral of the equation. The variable of integration on the left side is t while on the right side is $\vec{L}_{\text{tot}}$. For the limits of integration, at time t_i we assume that the angular momentum is $\vec{L}_{\text{tot.i}}$ and at time t_f the angular momentum is $\vec{L}_{\text{tot.f}}$:

$$\int_{t_i}^{t_f} \Sigma \vec{\tau}_{\text{ext}} dt = \int_{\vec{L}_{\text{tot.i}}}^{\vec{L}_{\text{tot.f}}} d\vec{L}_{\text{tot}}. \tag{5.36}$$

We now define the left-hand side of the equation to be equal to the rotational impulse $\vec{\mathcal{I}}_{\text{ext.r}}$ imparted by the net external torque during the time interval from t_i to t_f:

$$\vec{\mathcal{I}}_{\text{ext.r}} := \int_{t_i}^{t_f} \Sigma \vec{\tau}_{\text{ext}} dt. \tag{5.37}$$

Because $t_f > t_i$, the direction of the rotational impulse will be the same as that of the net external torque. With this definition in mind Equation (5.36) becomes:

$$\vec{\mathcal{I}}_{ext.r} = \int_{\vec{L}_{tot.i}}^{\vec{L}_{tot.f}} d\vec{L}_{tot} = \vec{L}_{tot.f} - \vec{L}_{tot.i} = \Delta \vec{L}_{tot}, \tag{5.38}$$

or

$$\vec{L}_{tot.i} + \vec{\mathcal{I}}_{ext.r} = \vec{L}_{tot.f}. \tag{5.39}$$

Equation (5.39) is the Angular Momentum-Impulse Theorem (AMIT), which is really Newton's Second Law for rotational motion written in a different form. Therefore, it is always true. It says that the final angular momentum of a system of objects is equal to whatever the total initial angular momentum was plus any additional angular momentum (impulse) that was imparted to the system by the net external torque. AMIT is, of course, also a vector equation.

If the rotational impulse $\vec{\mathcal{I}}_{ext.r}$ is zero (or small enough that we can approximate it to be zero), Equation (5.39) simplifies and we obtain the theorem of COAM:

$$\vec{L}_{tot.i} = \vec{L}_{tot.f}, \tag{5.40}$$

which states that the total angular momentum remains constant in both magnitude and direction.

There are again a couple of important points we need to make here. First, we repeat that while AMIT is always true, COAM is true only if the rotational impulse due to external torques is zero or negligible. Second, as we went from the most general expression of Newton's Second Law for rotational motion (which is always true) to AMIT, we did not require that the moment of inertia remain constant. Similarly, we did not require that the moment of inertia remain constant when we obtained COAM from AMIT. In other words, AMIT and COAM are useful to study systems in which the moment of inertia changes. On the other hand, $\Sigma \vec{\tau} = I\vec{\alpha}$ is only applicable if I is constant. This is in complete analogy to the relationship between MIT, COM, and $\Sigma \vec{F} = m\vec{a}$.

Example 3: Spinning ice skater. Explain why ice skaters spin faster when they tuck their arms in. The same explanation applies in diving, when high-divers make their bodies more compact as they flip through the air on their way down to the water.

These are examples where *internal* forces cause a redistribution of an object's (or system's) mass about the axis of rotation. This redistribution will cause the moment of inertia to change, since I depends on how the mass is distributed about the rotation axis. Given that there is no rotational impulse due to external torques, COAM can be applied:

$$\vec{L}_{\text{tot.i}} = \vec{L}_{\text{tot.f}}$$

$$\Rightarrow I_{\text{tot.i}}\vec{\omega}_i = I_{\text{tot.f}}\vec{\omega}_f$$

$$\Rightarrow \vec{\omega}_f = \frac{I_{\text{tot.i}}}{I_{\text{tot.f}}}\vec{\omega}_i, \tag{5.41}$$

meaning that a change in the moment of inertia will also cause the angular velocity to change.

The reason why there is no rotational impulse due to external torques is because both the ice skater and diver rotate about an axis going through their center of mass. Thus, the gravitational force does not produce a torque, neither does the reaction force from the ice on the skater, and so COAM can be applied. The ice skater can spread their arms out or tuck them in to change the moment of inertia about the axis through their center of mass. The closer their arms are to their body, the smaller the moment of inertia. Thus, when their moment of inertia decreases ($I_{\text{tot.f}} < I_{\text{tot.i}}$), their angular velocity has to increase for the angular momentum to remain constant. So, they spin faster!

Example 4: A tethered mass. Figure 5.6 shows a small mass m that is moving in a circle on a smooth horizontal surface while attached to a massless, inelastic string. The circular trajectory has radius R_1 and the mass has velocity $\vec{v}_1$ tangent to the circle. The string goes through a small hole that is located at the circle's center O so that we can exert a vertical force $\vec{F}_1$ at the string's end A, as shown. It is due to this force that the radius of the trajectory remains constant. At some point, we increase the magnitude of the force to $F_2 > F_1$ causing the string's end A to move downwards and m to move around a circle of radius $R_2 < R_1$. What is the magnitude of the linear velocity $\vec{v}_2$ with which m is now moving?

First, we note that the axis of rotation z for this mass is at the center of the circle O, perpendicular to the plane of the trajectory. Now, let's look at the forces exerted on m, shown in Figure 5.6. The gravitational force $\vec{F}_g$ and the reaction force $\vec{n}$ from the horizontal surface are parallel to the axis of rotation and, thus, produce no torque about this axis.

In addition, there is a force due to tension exerted on the mass. Since the rope is inelastic, this force is the same in magnitude as the force exerted on the string's end A. Thus, at time t_1 when the exerted force is $\vec{F}_1$, we have that $F_{T_1} = F_1$ (Figure 5.6a). Similarly, at time t_2 we have that $F_{T_2} = F_2$ (Figure 5.6b). The force due to tension is the only force that has a component that points toward the center of the circle and so it plays the role of the centripetal force, allowing the mass to move in a circle. However, since the force due to tension

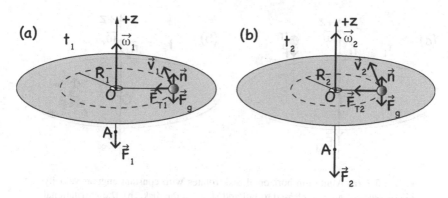

Figure 5.6 (a) A mass is executing circular motion on a smooth horizontal plane with the help of a vertical force $\vec{F}_1$ that is exerted on it via a string. There are three forces exerted on the mass: The gravitational force $\vec{F}_g$, the reaction force $\vec{n}$, and the tension $\vec{F}_{T_1}$. The tension plays the role of the centripetal force and since the string is inelastic, $F_{T_1} = F_1$. (b) When the force on the string becomes $\vec{F}_2$ where $F_2 > F_1$, the radius of the circle the mass is transcribing decreases ($R_2 < R_1$).

points toward the center of the circle, it is antiparallel to its position vector. Therefore, this force does not produce any torque either. Overall, since the forces external to the mass $\vec{F}_g$, $\vec{n}$, and $\vec{F}_T$ do not produce torque, we can use COAM.

We will solve this problem in two different ways, so that it becomes explicit that the magnitude of the angular momentum of a point mass as it moves around a circle can be expressed via Equation (5.8) ($L = rmv$) or Equation (5.14) ($L = I\omega$), where $I = mr^2$ for a point mass. Using the coordinate system defined in Figure 5.6 and that the mass is rotating counterclockwise, so that its angular momentum is along the $+z$ axis, we have:

First method: We use Equation (5.8) and apply COAM

$$\vec{L}_{\text{tot.1}} = \vec{L}_{\text{tot.2}}$$
$$\Rightarrow +L_{\text{tot.1}} = +L_{\text{tot.2}}$$
$$\Rightarrow R_1 m v_1 = R_2 m v_2$$
$$\Rightarrow v_2 = \frac{v_1 R_1}{R_2}. \tag{5.42}$$

Second method: We use Equation (5.14) and apply COAM

$$\vec{L}_{\text{tot.1}} = \vec{L}_{\text{tot.2}}$$
$$\Rightarrow +L_{\text{tot.1}} = +L_{\text{tot.2}}$$
$$\Rightarrow mR_1^2 \omega_1 = mR_2^2 \omega_2. \tag{5.43}$$

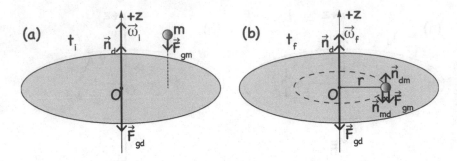

Figure 5.7 (a) A uniform horizontal disk rotates with constant angular velocity $\vec{\omega}_i$. A piece of gum is released to fall and stick on the disk. (b) The gravitational forces $\vec{F}_{gm}$ and $\vec{F}_{gd}$ and the reaction force from the table on the disk $\vec{n}_d$ are external forces. The reaction forces $\vec{n}_{md}$ on the disk by m and $\vec{n}_{dm}$ on m by the disk are internal forces and cancel out for the system. None of these forces produce a torque about z, and thus COAM can be applied to the system.

But, we know that $v = \omega r$. Therefore, our previous result becomes:

$$mR_1 v_1 = mR_2 v_2 \Rightarrow v_2 = \frac{v_1 R_1}{R_2}. \tag{5.44}$$

We see that the two definitions for the angular momentum of a point mass when it is executing circular motion give the same result, as they should!

Example 5: Rotating disk and bubble gum. The uniform horizontal disk shown in Figure 5.7 is rotating without friction about a vertical axis z that goes through its center O and is perpendicular to its plane. The disk has initial angular velocity $\vec{\omega}_i$ with direction as shown. The disk's moment of inertia about the z axis is I_d. A small piece of bubble gum of mass m is released from rest from above the disk. The piece of gum sticks to the disk at a distance r from the axis of rotation. Find the angular velocity of the disk-gum system after the bubble gum sticks to the disk.

Here we have a collision between the disk and the bubble gum. Figure 5.7a shows the instant right before the collision (i.e., initial), while Figure 5.7b shows the instant right after the collision (i.e., final). The coordinate system and all the forces exerted on the two objects that comprise the disk-gum system are shown for these two times. The collision is perfectly inelastic since, in the end, both masses stick together and move with the same angular velocity.

During the collision between the two objects, there are external forces. These include the gravitational forces $\vec{F}_{gm}$ and $\vec{F}_{gd}$ on the gum and disk, respectively. Furthermore, there is a reaction force $\vec{n}_d$ from the surface (i.e., a table) that supports the disk, or the disk would fall. However, none of these forces produce torque because they are all parallel to the axis of rotation z. The reaction force $\vec{n}_{md}$ on the disk by m and the reaction force $\vec{n}_{dm}$ on m by the disk are internal and they cancel out for the system. Overall, we see that $\Sigma\vec{\tau}_{ext} = 0$. Thus, COAM can be applied. We have:

$$\vec{L}_{tot.i} = \vec{L}_{tot.f}$$

$$\Rightarrow \vec{L}_{d.i} + \underbrace{\vec{L}_{m.i}}_{0} = \vec{L}_{(d+m).f}. \qquad (5.45)$$

Initially $\vec{L}_{m.i} = 0$ because the gum was moving parallel to the z axis before sticking to the disk, and so there is no angular momentum about this axis.

Equation (5.45) is a vector equation. Since we know that the disk's initial angular momentum points in the $+z$ direction, COAM dictates that the total angular momentum after the collision must also point in the $+z$ direction. Then we have:

$$+L_{d.i} = +L_{(d+m).f}$$

$$\Rightarrow I_d\omega_i = (I_d + I_m)\omega_f. \qquad (5.46)$$

When the gum sticks on the disk, we treat it as a point mass rotating about the z axis. Thus, its moment of inertia is $I_m = mr^2$, and we have:

$$\omega_f = \frac{I_d\omega_i}{I_d + I_m} = \frac{I_d\omega_i}{I_d + mr^2}. \qquad (5.47)$$

Since $I_d < I_d + mr^2$ we see that $\omega_f < \omega_i$ and, thus, the disk will slow down.

Exercises

(i) In Example 2, we computed the angular momentum at time $t = 0$. Considering the same system, (a) compute the angular momentum as a function of time, (b) find the net torque on the mass at time $t = 0$, and (c) compare the net torque found in part (b) to the result of Example 2 from Chapter 2.

(ii) A uniform pulley of radius R and mass M is free to rotate (no friction) about a horizontal axis that goes through its center O and is perpendicular to the plane of the pulley. A massless and inelastic rope is wound around the pulley at its circumference and a mass m is hanging from the rope's free end. This system is shown in Figure 5.8. Assuming that the system is released from rest at time $t_0 = 0$ and that the rope is not slipping against the rim of the pulley, find (a) the acceleration of the mass m, (b) the rate

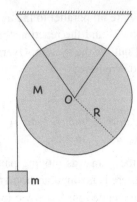

Figure 5.8 A uniform pulley is free to rotate about a horizontal axis that goes through its center O and is perpendicular to the plane of the pulley. A massless and inelastic rope is wound around the pulley at its circumference and a mass m is hanging from the rope's free end.

of change of the angular momentum of the system pulley-mass about the pulley's axis of rotation, (c) the change in angular momentum between time $t = t_1$ and time $t = 2t_1$, and (d) the angular momentum of the system about the pulley's axis of rotation the instant t_1 when the mass m has velocity $\vec{v}_1$. You are given the gravitational acceleration g and that the moment of inertia of the pulley about its axis of rotation through O is $I_O = \frac{1}{2}MR^2$.

(iii) Two uniform horizontal disks D_1 and D_2 have moments of inertia I_1 and I_2, respectively. They are rotating about a common axis z that goes through their centers and is perpendicular to their planes. The disks are a small vertical distance apart and are rotating in opposite directions with angular velocities of magnitude ω_1 and ω_2, respectively. By exerting an upward force on D_2, we move the disk up so that it comes in contact with D_1 and the two disks stick together. Find the magnitude and direction of the angular velocity of the system after the disks stick together.

(iv) A carousel of radius $R = 4.8$ m is at rest but free to rotate without friction about an axis perpendicular to its plane that goes through its center. Its moment of inertia about this axis is $I_{car} = 75$ kgm^2. A child of mass $m = 40$ kg is running at $v = 5.0$ m/s in a direction tangent to the rim of the carousel. Suddenly, the child jumps, lands on the rim of the carousel, and immediately stops (with respect to the carousel). Find the resulting angular velocity of the system.

(v) A circus clown of mass $m = 100$ kg stands at point P on the outer edge of a large uniform carousel of radius $R = 20$ m and mass $M = 2,000$ kg. The carousel can rotate without friction about an axis that goes through its center O and is perpendicular to its plane. The carousel and clown are initially at rest. At time $t = 0$ the clown starts running without slipping around the edge of the disk at a speed of $v = 2.0$ m/s with respect to the ground and in a direction that is perpendicular to the radius OP. Find (a) the carousel's angular momentum, (b) the carousel's angular velocity, and (c) the linear velocity of any point of the rim (e.g., point P) of the carousel. The moment of inertia of the carousel about its axis of rotation is $I_{car} = \frac{1}{2}MR^2$

6
Work and Energy

Now that we have learned about angular momentum and its conservation, it is time to talk about energy and its conservation in the case of rotational motion. This is the last topic we will cover before we can say that we have seen how all the concepts we learned in linear mechanics apply to rotational mechanics.

6.1 Review: Work due to a Force

When we first encountered the concept of energy in linear motion, we learned that energy is hard to define. We instead started by defining the work done by a force:

$$W = \int dW = \int_{S_i}^{S_f} \vec{F} \cdot d\vec{S} = \int_{S_i}^{S_f} F \, dS \cos \phi, \tag{6.1}$$

where $\vec{F}$ is the force exerted on the object, $d\vec{S}$ is the infinitesimal displacement of the object, and ϕ is the angle between these two vectors. dW is therefore the work that this force does on the object over this displacement. The SI unit of work is the Joule [J] = [Nm]. Summing up all the infinitesimal amounts of work as the object goes from location S_i to S_f will give the total work W done by force $\vec{F}$ during this motion. We had discussed in introductory physics that since this is a scalar (i.e., dot) product between force and displacement, only the component of the force that is parallel (or antiparallel) to the displacement does work. If both F and $\cos \phi$ remain constant during the motion, then $W = F \Delta S \cos \phi$.

We also learned in introductory physics that work expresses how much energy is transferred from one object to another (energy of object 1 is transferred via the work of a force to object 2) or how much energy is transformed from one form to another (energy of form X is transformed via the work of a force to energy of form Y). Then, using the concept of work and Newton's

Second Law, we defined kinetic energy and derived the Work-Kinetic Energy theorem (WKET).

We will follow a similar trajectory here to show that the reasoning behind all the derivations in this chapter remains the same as our reasoning when we were studying linear motion. So, off we go to define the concept of work in rotational motion!

6.2 Work due to a Torque

Figure 6.1 shows a rigid, extended, and irregular body (imagine a potato) of mass M, where M is the integral over all infinitesimal masses which make up the body $\left(M = \int dm\right)$. The body can rotate about a fixed axis that goes through point O and is perpendicular to the plane of this page. An external force $\vec{F}$ is applied to the object at point P and rotates it about this axis. Let's assume $\vec{F}$ is horizontal and in the plane of the page for simplicity. If the body rotates for an infinitesimal time interval dt, P transcribes an angle $d\theta$ and travels an arc length dS. Because the arc length is infinitesimal, it is equal to the magnitude of the displacement vector $d\vec{S}$. Since this is a rigid object, all points of this object will rotate by the same $d\theta$.

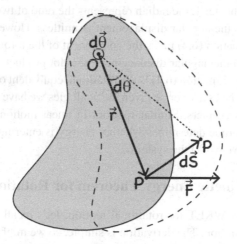

Figure 6.1 Due to a force $\vec{F}$ exerted at point P, the rigid body rotates about a fixed axis through point O and perpendicular to the plane of the page. After an infinitesimal amount of time dt, the object rotates by an angle $d\theta$, and point P travels an arc length dS. Because of the infinitesimal motion, the arc length dS is equal to the magnitude of the displacement vector $d\vec{S}$. $d\vec{\theta}$ is directed along the axis of rotation and, because of the counterclockwise rotation, points out of the page.

We derived in Chapter 1 that $d\vec{S} = d\vec{\theta} \times \vec{r}$, where $|\vec{r}| = r$ is the radius of the circle P transcribes and $\vec{r}$ goes from the axis of rotation to the point P where the force is exerted. Using Equation (6.1), the work dW done by the force $\vec{F}$ during the interval dt is given by:

$$dW = \vec{F} \cdot d\vec{S} = \vec{F} \cdot (d\vec{\theta} \times \vec{r}). \tag{6.2}$$

We can now use the identity that $\vec{A} \cdot (\vec{B} \times \vec{C}) = \vec{B} \cdot (\vec{C} \times \vec{A})$ so that Equation (6.2) becomes:

$$dW = d\vec{\theta} \cdot (\vec{r} \times \vec{F}) = d\vec{\theta} \cdot \vec{\tau} = \vec{\tau} \cdot d\vec{\theta}. \tag{6.3}$$

Therefore, we have obtained the infinitesimal work dW done during the time interval dt by the torque due to a force. If we sum up all the infinitesimal amounts of work as the object goes from angle θ_i to θ_f, we will get the total work W done by the torque $\vec{\tau}$ of the force $\vec{F}$ during the rotational motion:

$$W = \int dW = \int_{\theta_i}^{\theta_f} \vec{\tau} \cdot d\vec{\theta} = \int_{\theta_i}^{\theta_f} \tau d\theta \cos\phi, \tag{6.4}$$

where ϕ is the angle between the two vectors $\vec{\tau}$ and $d\vec{\theta}$. If both τ and $\cos\phi$ remain constant during the motion, then $W = \tau \Delta\theta \cos\phi$. The SI unit of work in this definition is again the Joule (as it should be). This is because torque [Nm] is multiplied with angular displacement [rad] through a dot (scalar) product. As we discussed in Chapter 1, the radian represents the ratio of two quantities with the same units, so the angular displacement is unitless. However, because the dot product in Equation (6.4) picks the component of the torque that is parallel (or antiparallel) to the angular displacement, the dot product of [Nm·rad] will represent the Joule. Equation (6.4) is the rotational equivalent of Equation (6.1), and should not come as a surprise given the analogies we have discussed extensively in previous chapters. Similar to work in linear motion, the sign of the work done by a torque determines whether energy is entering (positive work) or leaving (negative work) the system.

6.3 Work-Kinetic Energy Theorem for Rotational Motion

Before we discuss WKET for rotational motion, let's recall how we derived WKET in linear motion. The derivation is simple, so we might as well do it!

Using our definition from Equation (6.1), we can see that the net work done on a mass m (due to the net force exerted on it) is:

$$W_{\text{net}} = \int_{S_i}^{S_f} \Sigma\vec{F} \cdot d\vec{S}. \tag{6.5}$$

If the mass of the object is constant, then we can use Newton's Second Law for constant mass $\Sigma\vec{F} = m\vec{a}$ to substitute for the net force. If we also remember that the acceleration is defined as $\vec{a} = d\vec{v}/dt$ and that the velocity is defined as $\vec{v} = d\vec{S}/dt$ we get:

$$W_{net} = \int_{S_i}^{S_f} m\vec{a}\cdot d\vec{S} = \int_{S_i}^{S_f} m\frac{d\vec{v}}{dt}\cdot d\vec{S} = \int_{S_i}^{S_f} md\vec{v}\cdot\frac{d\vec{S}}{dt} = \int_{v_i}^{v_f} md\vec{v}\cdot\vec{v}. \quad (6.6)$$

Through this process, we have changed our variable of integration to $\vec{v}$ which is why our limits of integration in the last expression have changed accordingly.

Now, we have assumed that the mass is constant, so it can come out of the integral. Furthermore, we note that $d(v^2) = d(\vec{v}\cdot\vec{v}) = d\vec{v}\cdot\vec{v} + \vec{v}\cdot d\vec{v} = 2d\vec{v}\cdot\vec{v}$. In other words, $d\vec{v}\cdot\vec{v} = \frac{1}{2}d(v^2)$. Then, our previous result becomes:

$$W_{net} = \frac{m}{2}\int_{v_i}^{v_f} d(v^2) = \frac{1}{2}mv_f^2 - \frac{1}{2}mv_i^2. \quad (6.7)$$

At this point in our study of linear motion, we had defined the kinetic energy (KE) of an object as:

$$KE = \frac{1}{2}mv^2, \quad (6.8)$$

which also has the SI unit of the Joule [kgm^2/s^2]=[J]. Using this definition, Equation (6.7) becomes what is known as WKET:

$$W_{net} = KE_f - KE_i \Rightarrow KE_i + W_{net} = KE_f. \quad (6.9)$$

WKET says that the net work, done by all forces, results in the change in the kinetic energy of the object. We will now derive WKET for rotational motion in a similar way.

Using our definition from Equation (6.4), we can see that the net work done by the net torque is:

$$W_{net.torque} = \int_{\theta_i}^{\theta_f} \Sigma\vec{\tau}\cdot d\vec{\theta}. \quad (6.10)$$

If the moment of inertia is constant, then we can use Newton's Second Law for rotational motion for a constant moment of inertia $\Sigma\vec{\tau} = I\vec{\alpha}$ to substitute for the net torque. If we also remember that angular acceleration is defined as $\vec{\alpha} = d\vec{\omega}/dt$ and that the angular velocity is defined as $\vec{\omega} = d\vec{\theta}/dt$ we get:

$$W_{net.torque} = \int_{\theta_i}^{\theta_f} I\vec{\alpha}\cdot d\vec{\theta} = \int_{\theta_i}^{\theta_f} I\frac{d\vec{\omega}}{dt}\cdot d\vec{\theta} = \int_{\theta_i}^{\theta_f} Id\vec{\omega}\cdot\frac{d\vec{\theta}}{dt} = \int_{\omega_i}^{\omega_f} I\vec{\omega}\cdot d\vec{\omega}. \quad (6.11)$$

Again, through this process, we have changed our variable of integration to $\vec{\omega}$ which is why our limits of integration in the last expression have changed accordingly.

Now, we have assumed that the moment of inertia is constant, so it can come out of the integral. Furthermore, we note that $d(\omega^2) = d(\vec{\omega} \cdot \vec{\omega}) = d\vec{\omega} \cdot \vec{\omega} + \vec{\omega} \cdot d\vec{\omega} = 2d\vec{\omega} \cdot \vec{\omega}$. In other words, $d\vec{\omega} \cdot \vec{\omega} = \frac{1}{2}d(\omega^2)$. Then, our previous result becomes:

$$W_{\text{net.torque}} = \frac{I}{2} \int_{\omega_i}^{\omega_f} d(\omega^2) = \frac{1}{2}I\omega_f^2 - \frac{1}{2}I\omega_i^2. \qquad (6.12)$$

At this point, we define the rotational kinetic energy (KE_{rot}) of an object as:

$$KE_{\text{rot}} = \frac{1}{2}I\omega^2, \qquad (6.13)$$

which, just like the translational kinetic energy, has SI units of $[\text{kgm}^2/\text{s}^2] = [\text{J}]$. Using this definition, Equation (6.12) becomes what is known as WKET for rotational motion:

$$W_{\text{net.torque}} = KE_{\text{rot.f}} - KE_{\text{rot.i}} \Rightarrow KE_{\text{rot.i}} + W_{\text{net.torque}} = KE_{\text{rot.f}}. \qquad (6.14)$$

In this form, WKET says that the net work done by all torques results in the change in the rotational kinetic energy of the object.

Of course, a net force can result in both a linear acceleration and, via its torque, an angular acceleration. Therefore, the work of this net force can be responsible for a change in both the linear and the rotational kinetic energy. As a result, the most general expression of WKET can be obtained by adding Equations (6.9) for translational motion and (6.14) for rotational motion:

$$KE_i + KE_{\text{rot.i}} + (W_{\text{net.force}} + W_{\text{net.torque}}) = KE_f + KE_{\text{rot.f}}. \qquad (6.15)$$

6.3.1 WKET for a System of Objects

We have just derived our equations for $W_{\text{net.torque}}$ (Equation (6.10)) and WKET (Equation (6.14)) for a single object. However, we can easily extend these equations and apply them to a system of objects, where we could have multiple forces or torques on different objects in the system. How can we find $W_{\text{net.torque}}$ for the system? How does WKET apply to the system as a whole?

In linear motion, to find the total kinetic energy of a system of n objects at an arbitrary time, we simply add the individual kinetic energies of all the objects in the system. Similarly, for a system of n rotating objects, we find the system's total rotational kinetic energy at an arbitrary time by summing up the individual rotational kinetic energies at that time of all the objects in the system. That is:

$$KE_{\text{rot.tot}} = KE_{\text{rot.1}} + KE_{\text{rot.2}} + ... + KE_{\text{rot.n}}, \qquad (6.16)$$

where $KE_{rot.tot}$ is the total KE of the system, $KE_{rot.1}$ is the rotational KE of object 1, $KE_{rot.2}$ is the rotational KE of object 2, etc.

To find the net torque due to all forces on a system of objects, as we discussed for the case of Conservation of Angular Momentum, only the torques due to external forces matter. The internal torques cancel out due to Newton's third law. The net external torque provides work to change the system's rotational kinetic energy. The work done on the system by the net external torque is equal to the sum of the work provided by each external torque:

$$W_{net.torque} = W_{torque.1} + W_{torque.2} + ...$$ (6.17)

This is analogous to finding the net work done on a system of objects in linear motion via the sum of the work done by each external force. Though kinetic energy is always positive, because the work due to a torque can be positive or negative, it is important that we account for the correct signs in the terms of Equation (6.17).

6.4 Rotational Kinetic Energy of Systems of Discrete Masses and Continuous Rigid Bodies

In the previous section, we stumbled upon the definition of the rotational kinetic energy, $KE_{rot} = \frac{1}{2}I\omega^2$, while deriving WKET. In this section, we show how this expression comes about physically and why it represents the energy an object has because of its rotational motion. The derivation in this section will remind us of our derivations for the moment of inertia and angular momentum of systems of discrete masses and continuous rigid bodies in Chapters 4 and 5, respectively.

Once again, let's study an arbitrarily shaped rigid body, for example, the potato shown in Figure 6.2. This potato rotates with angular velocity $\vec{\omega}$ about a fixed axis z. We will first consider this rigid body as a collection of n small masses $m_1, m_2, \ldots, m_n$, whose position vectors and linear velocities are $\vec{r}_1, \vec{r}_2, \ldots, \vec{r}_n$, and $\vec{v}_1, \vec{v}_2, \ldots, \vec{v}_n$, respectively. Because the potato is a rigid body and the distance between any two small masses remains constant, it must be the case that every small mass rotates with the same angular velocity $\vec{\omega}_1 = \vec{\omega}_2 = \ldots = \vec{\omega}_n = \vec{\omega}$. We also know that the linear velocity of each mass is dependent on the distance between the mass and the axis of rotation, since $\vec{v} = \vec{\omega} \times \vec{r}$. Therefore, $\vec{v}_1 = \vec{\omega} \times \vec{r}_1, \vec{v}_2 = \vec{\omega} \times \vec{r}_2, \ldots, \vec{v}_n = \vec{\omega} \times \vec{r}_n$, and as discussed in Chapter 1, the magnitude of each small mass's linear velocity is given by $v_1 = \omega r_1, v_2 = \omega r_2, \ldots, v_n = \omega r_n$.

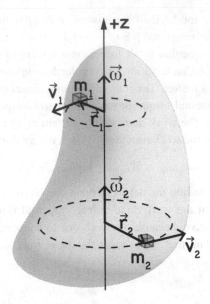

Figure 6.2 A potato of mass M represents any arbitrarily shaped rigid object. This
object is rotating about an axis z that remains fixed. The potato consists of many
small masses m_1, m_2, etc. whose position vectors and linear velocities are $\vec{r}_1$, $\vec{r}_2$,
etc., and $\vec{v}_1$, $\vec{v}_2$, etc., respectively. Since the potato is rigid, all small masses have
the same angular velocity $\vec{\omega}_1 = \vec{\omega}_2 = \ldots = \vec{\omega}$. The potato's rotational kinetic
energy can be shown to be $KE_{\text{rot}} = \frac{1}{2}I\omega^2$, where I is its moment of inertia.

The rigid body is not executing translational motion – it is simply rotating.
However, it is still in motion and, thus, there is some energy associated with that
motion. How do we find this energy? Since the potato is made of many small
masses, we can find its kinetic energy by summing up the kinetic energies of
these small masses. We have:

$$KE_{\text{tot}} = KE_1 + KE_2 + \ldots + KE_n$$

$$= \frac{1}{2}m_1 v_1^2 + \frac{1}{2}m_2 v_2^2 + \ldots + \frac{1}{2}m_n v_n^2. \tag{6.18}$$

We see that the kinetic energy due to rotation can be completely represented via
linear quantities. However, if we now substitute $v_1 = \omega r_1, v_2 = \omega r_2, \ldots, v_n =$
ωr_n for each small mass into Equation (6.18), we re-express the kinetic energy
due to the different linear speeds in terms of the common angular speed.
Because ω quantifies the rotational motion of all small masses which make
up the potato, and thus the potato as a whole, it can be used to represent the

rotational motion and, thus, the kinetic energy associated with the rigid body's rotation KE_{rot}:

$$KE_{rot} = \frac{1}{2}m_1\omega^2 r_1^2 + \frac{1}{2}m_2\omega^2 r_2^2 + ... + \frac{1}{2}m_n\omega^2 r_n^2$$

$$= \frac{1}{2}\omega^2(m_1 r_1^2 + m_2 r_2^2 + ... + m_n r_n^2)$$

$$= \frac{1}{2}\omega^2\left(\Sigma_{i=1}^{n} m_i r_i^2\right). \tag{6.19}$$

As we know, the last sum is the moment of inertia of a system of discrete small masses. However, the potato is really a continuous object. Therefore, we can make this sum more accurate by considering the limit where the masses become infinitesimal: dm_1, dm_2, ..., dm_n, so that the sum becomes an integral over m, and we are left with the rotational kinetic energy of a rigid body:

$$KE_{rot} = \frac{1}{2}\omega^2\left(\int r^2 dm\right) \Rightarrow KE_{rot} = \frac{1}{2}I\omega^2. \tag{6.20}$$

For a point mass rotating about a fixed axis, we can use either $\frac{1}{2}mv^2$ or $\frac{1}{2}I\omega^2$, where $I = mr^2$, to express its kinetic energy. As mentioned before, using angular quantities instead of linear quantities is just a more convenient approach to quantify rotational motion.

The result of Equation (6.20) is, again, somewhat expected. We could have easily found KE_{rot} by using the equation for $KE = \frac{1}{2}mv^2$ and substituting the rotational analogs I and ω for m and v, respectively. Although obtaining KE_{rot} in this manner would have been consistent with what we have said thus far about drawing analogies from our knowledge of translational motion, it would not have been a mathematically satisfying derivation.

Lastly, recall for the case of the translational kinetic energy that since the linear momentum has magnitude $p = mv$, the kinetic energy can be expressed as $KE = \frac{1}{2}mv^2 = \frac{p^2}{2m}$. Similarly, we can express the rotational kinetic energy in terms of the angular momentum. Since the magnitude of the angular momentum is given by $L = I\omega$, we have:

$$KE_{rot} = \frac{1}{2}I\omega^2 = \frac{1}{2}I\left(\frac{L}{I}\right)^2 \Rightarrow KE_{rot} = \frac{L^2}{2I}. \tag{6.21}$$

This should also not come as a surprise when considering the analogies that $\vec{p}$ is replaced by $\vec{L}$ while m is replaced by I when moving from linear to rotational motion.

6.5 Power due to a Force or a Torque

Before we talk about energy conservation, let's open a parenthesis and discuss the instantaneous rate at which a force does work (i.e., the instantaneous rate at which energy is being transferred or transformed by the force). In linear motion we had defined the instantaneous power $\mathcal{P}$ of a force to be the physical quantity that describes this rate. Mathematically, $\mathcal{P}$ is therefore defined as

$$\mathcal{P} = \frac{dW}{dt} = \frac{\vec{F} \cdot d\vec{S}}{dt} \Rightarrow \mathcal{P} = \vec{F} \cdot \vec{v}, \tag{6.22}$$

where we have used Equation (6.2) and the definition that $\vec{v} = d\vec{S}/dt$. Based on this definition, we see that the SI unit of power is the Watt: $[W]=[N\frac{m}{s}]=[\frac{J}{s}]$. If we are looking at the power due to the net force, then Equation (6.22) becomes:

$$\mathcal{P}_{net} = \Sigma\vec{F} \cdot \vec{v}. \tag{6.23}$$

However, we know from WKET (Equation (6.9)) that the work due to the net force is equal to the change in the translational kinetic energy. Therefore, we can take the derivative of WKET with respect to time as follows:

$$W_{net} = KE_f - KE_i$$

$$\Rightarrow \mathcal{P}_{net} = \frac{dW_{net}}{dt} = \frac{d(KE_f - KE_i)}{dt}. \tag{6.24}$$

By combining these last two results we see that:

$$\mathcal{P}_{net} = \frac{dW_{net}}{dt} = \frac{d(KE_f - KE_i)}{dt} = \Sigma\vec{F} \cdot \vec{v}. \tag{6.25}$$

In a similar way we can find the power delivered by a torque in rotational motion. Using Equation (6.3) and the definition that $\vec{\omega} = d\vec{\theta}/dt$ we get:

$$\mathcal{P}_{rot} = \frac{dW}{dt} = \frac{\vec{\tau} \cdot d\vec{\theta}}{dt} \Rightarrow \mathcal{P}_{rot} = \vec{\tau} \cdot \vec{\omega}, \tag{6.26}$$

which also has SI unit of the Watt: $[W]=[Nm\frac{rad}{s}]=[\frac{J}{s}]$. If we are looking at the power due to the net torque, then Equation (6.26) becomes:

$$\mathcal{P}_{net.torque} = \Sigma\vec{\tau} \cdot \vec{\omega}. \tag{6.27}$$

From WKET for rotational motion (Equation (6.14)) we know that the work due to the net torque is equal to the change in the rotational kinetic energy. Therefore, we can take the derivative of WKET with respect to time as follows:

$$W_{net.torque} = KE_{rot.f} - KE_{rot.i}$$

$$\Rightarrow \mathcal{P}_{net.torque} = \frac{dW_{net.torque}}{dt} = \frac{d(KE_{rot.f} - KE_{rot.i})}{dt}. \tag{6.28}$$

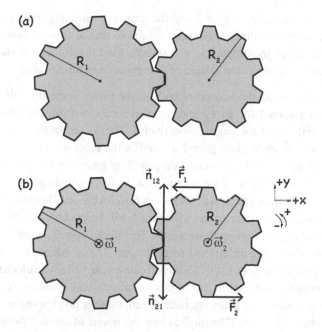

Figure 6.3 (a) Two gears of radii R_1 and R_2 are interlocked so that they rotate together without slipping. (b) Due to the force couple exerted on gear 2, the two gears start rotating in opposite directions. Only the force exerted by gear 2 on gear 1 ($\vec{n}_{21}$) is responsible for the rotation of gear 1. Because the gears are interlocked, they will have the same linear velocity at their points of contact and, thus, at all points of their circumferences.

By combining these last two results we see that

$$\mathcal{P}_{\text{net.torque}} = \frac{dW_{\text{net.torque}}}{dt} = \frac{d(KE_{\text{rot.f}} - KE_{\text{rot.i}})}{dt} = \Sigma\vec{\tau} \cdot \vec{\omega}. \qquad (6.29)$$

Again, given the analogies we have discussed between translational and rotational motion, the mathematical expressions for the power in rotational motion are not surprising. Just as with the work, the sign of the power determines whether energy is entering (positive power) or leaving (negative power) the system. Now let's look at two examples before we come back to study Conservation of Energy.

Example 1: Interlocking gears. The two uniform interlocking gears of Figure 6.3a are fixed on a table by axles that go through their centers and are perpendicular to their planes. Gear 1 and gear 2 have radii $R_1 = 1.0$ m and $R_2 = 0.5$ m, and moments of inertia $I_1 = 2.0$ kgm^2 and $I_2 = 1.0$ kgm^2, about their axles, respectively. Initially, the two gears are at rest. At time $t_0 = 0$ we start exerting a force couple on gear 2 and the gears start rotating without friction. This force couple results in a constant torque so that at time t_1 gear 1 has an angular velocity of magnitude $\omega_1 = 5.0$ rad/s. Find the amount of energy that the force couple gives to the system of two gears from time t_0 until time t_1.

Figure 6.3b shows the two gears with all the forces exerted on them, other than the gravitational forces $\vec{F}_g$ and reaction forces $\vec{N}$ from the table (which point into and out of the page, respectively). The forces shown are the force couple $\vec{F}_1$ and $\vec{F}_2$ exerted on gear 2 as well as the reaction forces $\vec{n}_{21}$ exerted on gear 1 by gear 2 and $\vec{n}_{12}$ exerted on gear 2 by gear 1. Why do the reaction forces have this direction? If we assume that the force couple on gear 2 has the direction shown, then gear 2 will rotate counterclockwise. This means that the tooth of gear 2 that is interlocked with gear 1 will be pushing down on gear 1, which is why $\vec{n}_{21}$ is pointing downwards. Therefore, from Newton's third law, gear 1 will be exerting an upward force on gear 2, $\vec{n}_{12}$. Also, since $\vec{n}_{21}$ is the only force exerted on gear 1, this force will cause gear 1 to rotate clockwise. For the system of two gears, these two forces are internal and, thus, they cancel out.

In addition, since the gears are interlocked, there is no slipping as one gear rotates against the other. This implies that the points of contact between gear 1 and gear 2 have the same linear speed. Since these points are at the gears' circumference, every point of the circumference of the gears will have the same linear speed. In other words:

$$v_1 = v_2 \Rightarrow \omega_1 R_1 = \omega_2 R_2 \Rightarrow \omega_2 = \omega_1 \frac{R_1}{R_2}. \tag{6.30}$$

Therefore, at time t_1 the angular speed of gear 2 is:

$$\omega_2 = 5.0 \text{ rad/s} \cdot \frac{1.0 \text{ m}}{0.5 \text{ m}} = 10 \text{ rad/s}. \tag{6.31}$$

To find the amount of energy given by the force couple (i.e., the work done by the two forces $W_{F_1 F_2}$), we apply WKET to the system from time t_0 to t_1:

$$\underbrace{KE_{\text{rot.1i}}}_{0} + \underbrace{KE_{\text{rot.2i}}}_{0} + W_{\text{net.torque}} = KE_{\text{rot.1f}} + KE_{\text{rot.2f}}. \tag{6.32}$$

The $KE_{\text{rot.i}}$ terms go to zero because the two gears start from rest. Now, for the net torque we need to account for all the forces:

$$\underbrace{W_{\tau F_{g_1}}}_{0} + \underbrace{W_{\tau F_{g_2}}}_{0} + \underbrace{W_{\tau N_1}}_{0} + \underbrace{W_{\tau N_2}}_{0} + W_{F_1 F_2} + \underbrace{W_{n_{12}} + W_{n_{21}}}_{\text{cancel out}} = KE_{\text{rot.1f}} + KE_{\text{rot.2f}}. \tag{6.33}$$

The first four work terms go to zero because the corresponding forces produce no torque, since these forces are exerted at the axis of rotation. In addition,

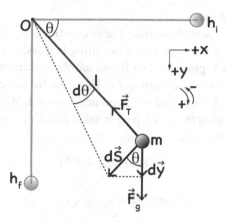

Figure 6.4 A simple pendulum consists of a massless inelastic string of length l fixed at point O and a point mass m attached to the free end of the string. The mass is released from rest from the horizontal position (at height h_i) and the motion is observed until it reaches the vertical position (at height h_f). Here it is also shown at some intermediate position.

because the two reaction forces $\vec{n}_{12}$ and $\vec{n}_{21}$ are internal and cancel out, their works cancel out. So, we are left with only the work due to the force couple:

$$W_{F_1 F_2} = \frac{1}{2} I_1 \omega_1^2 + \frac{1}{2} I_2 \omega_2^2$$
$$= \frac{1}{2} \cdot 2.0 \text{ kgm}^2 \cdot (5.0 \text{ rad/s})^2 + \frac{1}{2} \cdot 1.0 \text{ kgm}^2 \cdot (10 \text{ rad/s})^2 = 75 \text{ J}.$$

$$(6.34)$$

Example 2: The simple pendulum. The simple pendulum of Figure 6.4 consists of a massless and inelastic string of length l and a small (i.e., point) mass m. One end of the string is attached to the ceiling (at point O), and the mass is attached to the string's free end. The mass is released from rest from the horizontal position, which is at a height h_i above the ground. The pendulum's motion is observed until the string is vertical, at which point the mass m is at a height h_f above the ground. Show that you can apply WKET using either linear or angular quantities and obtain the same equation to describe the transformation of energy in the system as the mass goes from the initial height h_i to the final height h_f. You are given the acceleration due to gravity, g.

The goal of this example is to show that, in many cases, the motion of an object can be studied using linear or angular quantities. We should obtain the same result regardless of how we choose to parameterize the system.

Using linear quantities

First, let's study the pendulum using linear quantities. Figure 6.4 shows only the forces exerted on the mass (since the string is massless) when the pendulum is at an arbitrary position. The forces are the gravitational force $\vec{F}_g$ and the force due to tension of the string $\vec{F}_T$. Note that for our coordinate system in this example we have picked $+y$ to point downward. We are now ready to apply WKET (Equation (6.9)), keeping in mind that $KE_i = 0$ since the mass is released from rest. We have:

$$\underbrace{KE_i}_{0} + W_{\text{net}} = KE_f$$

$$\Rightarrow W_{F_g} + \underbrace{W_{F_T}}_{0} = KE_f$$

$$\Rightarrow W_{F_g} = \frac{1}{2}mv_f^2, \tag{6.35}$$

where we have used the fact that since the force of tension $\vec{F}_T$ is perpendicular to the displacement $d\vec{S}$ of the mass as it goes from h_i to h_f, it does no work on the mass.

To find the work done by $\vec{F}_g$, we first note that as the mass moves, the angle between the gravitational force $\vec{F}_g$ and the displacement $d\vec{S}$ changes. At the intermediate position shown in Figure 6.4, it is clear that the angle θ between the pendulum and the horizontal is the same as the angle between $d\vec{S}$ and $\vec{F}_g$. Therefore, to find the work done by $\vec{F}_g$, we integrate between $\theta_i = 0$ at the initial height h_i, and $\theta_f = \frac{\pi}{2}$ at the final height h_f:

$$W_{F_g} = \int_{h_i}^{h_f} \vec{F}_g \cdot d\vec{S} = \int_{h_i}^{h_f} F_g \, dS \cos\theta. \tag{6.36}$$

However, $\cos\theta = \frac{dy}{dS}$ and $dy = -dh$ (h decreases as y increases). Therefore,

$$W_{F_g} = \int_{h_i}^{h_f} F_g \, dS \frac{dy}{dS} = \int_{h_i}^{h_f} -F_g \, dh, \tag{6.37}$$

and Equation (6.35) becomes:

$$\int_{h_i}^{h_f} -F_g \, dh = \frac{1}{2}mv_f^2$$

$$\Rightarrow F_g(h_i - h_f) = \frac{1}{2}mv_f^2$$

$$\Rightarrow mgh_i - mgh_f = \frac{1}{2}mv_f^2. \tag{6.38}$$

Of course, the astute reader will observe this as simply Conservation of Mechanical Energy (discussed in the next section) applied to a pendulum.

Using angular quantities

Now let's use angular quantities to see if we can get the same result. To start, we note that the mass is executing circular motion about an axis of rotation that goes through point O on the ceiling and is perpendicular to the plane of this page. The radius of the circular trajectory is the length l of the string. Since this is a point mass, its moment of inertia about this axis is $I = mr^2 = ml^2$.

Let's apply WKET for rotational motion (Equation (6.14)), again keeping in mind that $KE_{rot.i} = 0$ since the mass is released from rest. We have:

$$\underbrace{KE_{rot.i}}_{0} + W_{net.torque} = KE_{rot.f}$$

$$\Rightarrow W_{\tau_{Fg}} + \underbrace{W_{F_T}}_{0} = KE_{rot.f}$$

$$\Rightarrow W_{\tau_{Fg}} = \frac{1}{2} I \omega_f^2, \tag{6.39}$$

where we have used the fact that since the force of tension $\vec{F}_T$ is always anti-parallel to the position vector as the mass goes from h_i to h_f, it does not produce a torque on the mass about the specified axis. Therefore, it does no work on the mass.

To find the work done by the torque of $\vec{F}_g$, we first note that as the mass moves, $\vec{\tau}_{F_g}$ and $d\vec{\theta}$ are parallel (they both point into the page in Figure 6.4). Therefore, using Equation (6.4), we have for the work due to $\vec{\tau}_{Fg}$:

$$W_{\tau_{Fg}} = \int_{\theta_i}^{\theta_f} \vec{\tau}_{F_g} \cdot d\vec{\theta} = \int_{\theta_i}^{\theta_f} \tau_{F_g} d\theta \cos 0 = \int_{\theta_i}^{\theta_f} \tau_{F_g} d\theta. \tag{6.40}$$

However, by definition, $\tau_{F_g} = F_g l \sin \phi = F_g l \sin \left(\frac{\pi}{2} - \theta \right) = F_g l \cos \theta$. Since $\theta_i = 0$ and $\theta_f = \frac{\pi}{2}$, our integral now becomes:

$$W_{\tau_{Fg}} = \int_0^{\frac{\pi}{2}} F_g l \cos \theta d\theta = F_g l [\sin \theta]_0^{\frac{\pi}{2}} = F_g l = F_g (h_i - h_f), \tag{6.41}$$

where the last substitution was made because we can see in Figure 6.4 that $h_i - l = h_f \Rightarrow l = h_i - h_f$. Finally, combining Equations (6.41) and (6.39) gives:

$$F_g (h_i - h_f) = \frac{1}{2} I \omega_f^2. \tag{6.42}$$

Since $I = ml^2$ and $v = \omega r = \omega l \Rightarrow \omega = \frac{v}{l}$, we have:

$$F_g(h_i - h_f) = \frac{1}{2}ml^2\frac{v_f^2}{l^2}$$

$$\Rightarrow mg(h_i - h_f) = \frac{1}{2}mv_f^2$$

$$\Rightarrow mgh_i - mgh_f = \frac{1}{2}mv_f^2. \qquad (6.43)$$

As expected, Equation (6.43), obtained by applying WKET for rotational motion, is the same as Equation (6.38), which was obtained by applying WKET for linear motion.

6.6 Conservation of Energy and Conservation of Mechanical Energy

When studying energy in translational motion in introductory physics, WKET was the starting point for deriving the theorems of Conservation of Energy (CE) and Conservation of Mechanical Energy (CME). Let's start this section by reviewing the derivations of CE and CME from WKET for translational motion. Our proof of the corresponding theorems for rotational motion will be similar.

We begin our derivation by recalling that all forces in nature can be divided into two categories: conservative forces (CFs) and nonconservative forces (NCFs). Therefore, the net force exerted on an object will be the sum of all the CFs and the NCFs. As a result, the work done by the net force, W_{net}, will be the sum of the work done by the CFs plus the work done by the NCFs:

$$W_{net} = W_{CF} + W_{NCF}. \qquad (6.44)$$

Therefore, we can re-write WKET (Equation (6.9)) as:

$$KE_i + W_{CF} + W_{NCF} = KE_f. \qquad (6.45)$$

The benefit of separating forces into conservative and nonconservative is that CFs have the property that their work is path-independent. Thus, the work of any CF can be written in terms of the change in the corresponding potential energy (U). Specifically:

$$W_{CF} = -\Delta U = -(U_f - U_i) = U_i - U_f, \qquad (6.46)$$

where subscripts i and f correspond to the initial and final points of the path along which the force does work, respectively. Substituting this expression for W_{CF} in Equation (6.45) gives:

$$KE_i + (U_i - U_f) + W_{\text{NCF}} = KE_f$$
$$\Rightarrow KE_i + U_i + W_{\text{NCF}} = KE_f + U_f. \qquad (6.47)$$

This result is, of course, CE! We see that WKET and CE are equivalent – one comes from the other via two substitutions. Therefore, we can use either WKET or CE to study the transfer and transformation of energy in physical systems. At times, one expression may be more advantageous to use than the other, depending on what we know about our physical system and what we are trying to find.

In our studies, we may encounter a system on which (a) there are no NCFs exerted (i.e., there are only CFs), or (b) there are NCFs but they cancel each other out so that there is no net NCF (i.e., $\Sigma \vec{F}_{\text{NCF}} = 0$), or (c) there is a net NCF but that force does not do any work (because it is perpendicular to the displacement, for example). In all cases, $W_{\text{NCF}} = 0$, and Equation (6.47) for CE becomes CME:

$$KE_i + U_i = KE_f + U_f. \qquad (6.48)$$

In other words, only the work done by NCFs changes the mechanical energy of a system (hence the name non-conservative!).

As we have seen throughout this book, there are situations in which it is more convenient to study a physical system in terms of angular quantities rather than linear quantities. So, let's derive CE and CME in terms of our angular quantities.

Similar to when we started the derivations for CE and CME in linear motion, let's start with WKET for rotational motion, and write the net work in terms of the torque due to CFs and NCFs:

$$W_{\text{net.torque}} = W_{\tau_{\text{CF}}} + W_{\tau_{\text{NCF}}}. \qquad (6.49)$$

Therefore, we can re-write WKET (Equation (6.14)) as:

$$KE_{\text{rot.i}} + W_{\tau_{\text{CF}}} + W_{\tau_{\text{NCF}}} = KE_{\text{rot.f}}. \qquad (6.50)$$

Again, the work done by the torque of a CF will lead to a potential energy (as seen in Example 2 with the simple pendulum). Thus, we have from Equation (6.46):

$$KE_{\text{rot.i}} + (U_i - U_f) + W_{\tau_{\text{NCF}}} = KE_{\text{rot.f}}$$
$$\Rightarrow KE_{\text{rot.i}} + U_i + W_{\tau_{\text{NCF}}} = KE_{\text{rot.f}} + U_f. \qquad (6.51)$$

This result is CE for rotational motion. Just as in translational motion, WKET and CE are equivalent.

In our studies, we may encounter a system on which (a) there are no torques due to NCFs (i.e., there are only torques due to CFs), or (b) there are torques due to NCFs but they cancel each other out so that there is no net torque due to NCFs (i.e., $\Sigma \vec{\tau}_{\text{NCF}} = 0$), or (c) there is a net torque due to NCFs but that torque does not do any work about the specified axis, for example. In all cases the work of the torque due to NCFs is zero, $W_{\tau_{\text{NCF}}} = 0$, and Equation (6.51) for CE becomes CME:

$$KE_{\text{rot.i}} + U_i = KE_{\text{rot.f}} + U_f. \tag{6.52}$$

It may seem strange that in Equations (6.51) and (6.52) (the rotational analogs for CE and CME, respectively), we have not specified "rotational" potential energy, as we have for the kinetic energies. This is because a system's potential energy depends only on its orientation or position in space. Whether we express this system's change in orientation or position in terms of linear or angular quantities will not affect the change in the potential energy associated with a CF. This is what we saw above in Example 2, where we found that the work done by the gravitational force and the work done by the torque of the gravitational force both correspond to the same change in the gravitational potential energy.

In some cases, an object will both translate and rotate. In such cases, we will need to apply WKET or CE for both the translational and rotational components of its motion to fully understand the behavior of the system. We will further discuss this in Chapter 7.

Example 3: Two masses and a pulley. A uniform pulley has mass $M = 4.0\,\text{kg}$ and radius $r = 1.0\,\text{m}$. The pulley is firmly fixed to a ceiling via an axle and can rotate with no friction about a fixed horizontal axis that goes through its center O and is perpendicular to its plane. A massless and inelastic string is wrapped around the pulley's circumference so that each end is hanging on either side of the pulley. A small mass $m_1 = 2.0\,\text{kg}$ is attached on one end while another small but heavier mass $m_2 = 4.0\,\text{kg}$ is attached to the other end. The configuration is shown in Figure 6.5a. At time $t_i = 0$, when the two masses m_1 and m_2 are at the same height, we release the system from rest. Find the magnitude of the pulley's angular velocity at time t_f when the two masses are separated by a distance $h = 2.0\,\text{m}$. Assume that the string does not slip against the pulley, that $g = 10\,\text{m/s}^2$, and that the pulley's moment of inertia about the specified axis of rotation is $I = \frac{1}{2}Mr^2$.

We start by noting all the forces exerted on the system of three masses (m_1, m_2, and the pulley). The forces exerted on the pulley are the gravitational force $\vec{F}_g$, the reaction force $\vec{R}$ from the ceiling via the axle, and the forces due to

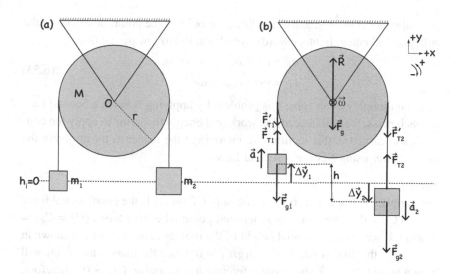

Figure 6.5 (a) Two masses m_1 and m_2 are hanging on either side of a pulley via a massless and inelastic string. Initially, the two masses are at the same height $h_i = 0$. The system is released from rest from this position. (b) Since m_2 is heavier than m_1 the pulley rotates clockwise when the system is released, increasing the distance h between m_1 and m_2. The forces exerted on the two masses and the pulley are the forces due to the tension on the string, the gravitational forces, and the reaction force $\vec{R}$ from the ceiling on the pulley via the axle.

tension $\vec{F}'_{T_1}$ and $\vec{F}'_{T_2}$. We note that since $\vec{F}_g$, $\vec{F}'_{T_1}$ and $\vec{F}'_{T_2}$ all point downward, $\vec{R}$ has to be upward for the pulley to be in equilibrium along the y axis. The forces exerted on the two masses m_1 and m_2 are the gravitational forces $\vec{F}_{g_1}$ and $\vec{F}_{g_2}$ and the forces due to tension $\vec{F}_{T_1}$ and $\vec{F}_{T_2}$, respectively. All forces are shown in Figure 6.5b. In addition, since $m_2 > m_1$, m_2 will move downward, m_1 will move upwards, and the pulley will rotate clockwise.

The fact that the string is inelastic and does not slip against the pulley allows us to easily relate the motion of all objects in the system. Since the string is inelastic, the magnitude of m_2's downward displacement is equal to the magnitude of m_1's upward displacement, that is, $\Delta y_1 = \Delta y_2$. Because the string is not slipping, these in turn are equal to the arc length ΔS by which the pulley rotates. That is, $\Delta y_1 = \Delta y_2 = \Delta S = r\Delta\theta$, where $\Delta\theta$ is the pulley's angular displacement. We also know that since $\Delta y_1 + \Delta y_2 = h = 2.0$ m at time t_f, then $\Delta y_1 = \Delta y_2 = r\Delta\theta = \frac{h}{2} = 1.0$ m. It is possible to prove this relation using 1D kinematics, but we will save that for the reader to do.

Applying both conditions (the string is inelastic and does not slip) means that the magnitudes of the accelerations of m_1, m_2, and of the linear acceleration of the points at the rim of the pulley are all equal. Similarly, the magnitudes of the velocities of m_1, m_2, and of the linear velocity of the points at the rim of the pulley are all equal. In other words, we obtain the following relations:

$$a_1 = a_2 = a_{\text{rim}} = \alpha r = a,$$

$$v_1 = v_2 = v_{\text{rim}} = \omega r = v. \qquad (6.53)$$

Even though we can solve this problem by applying Newton's Second Law for each mass, we will instead use work and energy theorems to apply the concepts we learned in this section. We encourage the reader to try to obtain the same answer using Newton's Second Law.

For m_1

Let's apply CE (Equation (6.47)). The only CF on m_1 is the gravitational force $\vec{F}_{g1}$, so CE will contain only gravitational potential energy terms ($U_1 = U_{g1} = m_1 g h_1$). If we define the initial height of the masses to be at $h = 0$, as shown in Figure 6.5, this means that $U_{i1} = m_1 g h_{i1} = 0$. Once the masses move, m_1 will be at a height $\Delta y_1 = \frac{h}{2}$ above our reference $h = 0$, and so $U_{f1} > 0$. Therefore, CE takes the form:

$$\underbrace{KE_{i1}}_{0} + \underbrace{U_{i1}}_{0} + W_{\text{NCF}_1} = KE_{f1} + U_{f1}$$

$$\Rightarrow W_{FT_1} = \frac{1}{2}m_1 v^2 + m_1 g \Delta y_1$$

$$\Rightarrow F_{T_1} \Delta y_1 \cos 0 = \frac{1}{2}m_1 v^2 + m_1 g \Delta y_1$$

$$\Rightarrow F_{T_1} \frac{h}{2} = \frac{1}{2}m_1 v^2 + m_1 g \frac{h}{2}. \qquad (6.54)$$

To obtain this equation we have also used the fact that $\vec{F}_{T_1}$ is constant and parallel to the displacement $\Delta \vec{y}_1$ to find the work W_{FT_1}.

For m_2

Let's again apply CE (Equation (6.47)). The only CF on m_2 is the gravitational force $\vec{F}_{g2}$, so CE will contain only gravitational potential energy terms ($U_2 = U_{g2} = m_2 g h_2$). Since we have defined the initial height of the masses to be at $h = 0$, this means that $U_{i2} = m_2 g h_{i2} = 0$. Once the masses move, m_2 will be at a height $\Delta y_2 = \frac{h}{2}$ below our reference $h = 0$, and so $U_{f2} < 0$. Therefore, CE takes the form:

$$\underbrace{KE_{i2}}_{0} + \underbrace{U_{i2}}_{0} + W_{\text{NCF}_2} = KE_{f2} + U_{f2}$$

$$\Rightarrow W_{FT_2} = \frac{1}{2}m_2 v^2 - m_2 g \Delta y_2$$

$$\Rightarrow F_{T_2} \Delta y_2 \cos \pi = \frac{1}{2} m_2 v^2 - m_2 g \Delta y_2$$

$$\Rightarrow -F_{T_2} \frac{h}{2} = \frac{1}{2} m_2 v^2 - m_2 g \frac{h}{2}. \tag{6.55}$$

To obtain this equation we have also used the fact that $\vec{F}_{T_2}$ is constant and anti-parallel to the displacement $\Delta \vec{y}_2$ to find the work $W_{F_{T_2}}$.

For the pulley M

Lastly, let's apply CE for the pulley (Equation (6.51)). The only CF on the pulley is the gravitational force $\vec{F}_g$, but since the pulley does not translate, its gravitational potential energy does not change. Thus $U_f = U_i$ and these terms will cancel out. Therefore, CE for the pulley takes the form:

$$\underbrace{KE_{\text{rot.i}}}_{0} + U_i + W_{\tau_{\text{NCF}}} = KE_{\text{rot.f}} + U_f \Rightarrow W_{\tau_{\text{NCF}}} = KE_{\text{rot.f}}$$

$$\Rightarrow W_{\tau_{F'T1}} + W_{\tau_{F'T2}} + \underbrace{W_{\tau_R}}_{0} = \frac{1}{2} I \omega^2 = \frac{1}{2} \left(\frac{1}{2} M r^2 \right) \frac{v^2}{r^2}$$

$$\Rightarrow W_{\tau_{F'T1}} + W_{\tau_{F'T2}} = \frac{1}{4} M v^2. \tag{6.56}$$

We have also used the fact that the force $\vec{R}$ does no work since it is exerted at the axis of rotation and thus does not produce a torque about this axis.

Now let's look at the work due to the torques. Since the pulley is rotating clockwise, $\Delta \vec{\theta}$ points into the page. However, since $\vec{\tau}_{F'_{T1}}$ points out of the page, the torque and the angular displacement are anti-parallel. On the other hand, $\vec{\tau}_{F'_{T2}}$ points into the page, so it is parallel to the displacement. Therefore, we have for their works:

$$W_{\tau_{F'T1}} = \tau_{F'_{T1}} \Delta \theta \cos \pi = -\tau_{F'_{T1}} \Delta \theta = -\left(F'_{T_1} r \sin \left(\frac{\pi}{2} \right) \right) \Delta \theta = -F'_{T_1} r \Delta \theta = -F'_{T_1} \frac{h}{2},$$

$$W_{\tau_{F'T2}} = \tau_{F'_{T2}} \Delta \theta \cos 0 = \tau_{F'_{T2}} \Delta \theta = \left(F'_{T_2} r \sin \left(\frac{\pi}{2} \right) \right) \Delta \theta = F'_{T_2} r \Delta \theta = F'_{T_2} \frac{h}{2}. \tag{6.57}$$

Here we have used $r \Delta \theta = \Delta S = h/2$ (as mentioned earlier) and, since the torques are constant, that $\tau_F = Fr \sin \phi$. Combining Equations (6.56) and (6.57) gives:

$$-F'_{T_1} \frac{h}{2} + F'_{T_2} \frac{h}{2} = \frac{1}{4} M v^2. \tag{6.58}$$

Let's make a quick observation: Equation (6.58) shows that, in contrast to a massless pulley, the two tension forces $\vec{F}'_{T_1}$ and $\vec{F}'_{T_2}$ cannot be equal if the pulley is to begin rotating. If they were equal, then the left-hand side of this equation would be zero, and since the right-hand side represents the pulley's kinetic energy, $KE_{rot.f} = 0$. Since the pulley starts from rest, this implies that the pulley would not rotate.

We will now add the three CE equations together (Equations (6.54), (6.55), (6.58)) to get the CE for the system of three masses, keeping in mind that $F_{T_1} = F'_{T_1}$ and $F_{T_2} = F'_{T_2}$ since they are action-reaction pairs:

$$F_{T_1}\frac{h}{2} - F_{T_2}\frac{h}{2} - F'_{T_1}\frac{h}{2} + F'_{T_2}\frac{h}{2} = \frac{1}{2}m_1v^2 + m_1g\frac{h}{2} + \frac{1}{2}m_2v^2 - m_2g\frac{h}{2} + \frac{1}{4}Mv^2$$

$$\Rightarrow 0 = g\frac{h}{2}(m_1 - m_2) + \frac{v^2}{2}\left(m_1 + m_2 + \frac{1}{2}M\right).$$

(6.59)

We see that all the works have canceled out – this makes sense since the forces due to tension are internal forces to the system. Therefore, when we look at the system as a whole, these forces cancel out so the net work done by these forces is zero. We can now continue and solve for v^2:

$$-gh(m_1 - m_2) = v^2\left(m_1 + m_2 + \frac{1}{2}M\right) \Rightarrow v^2 = \frac{-gh(m_1 - m_2)}{m_1 + m_2 + \frac{1}{2}M}, \quad (6.60)$$

and plug in the values to find v:

$$v^2 = \frac{-10 \text{ m/s}^2 \cdot 2.0 \text{ m}(2.0 \text{ kg} - 4.0 \text{ kg})}{2.0 \text{ kg} + 4.0 \text{ kg} + \frac{4.0 \text{ kg}}{2}} = 5.0\frac{\text{m}^2}{\text{s}^2} \Rightarrow v = \sqrt{5.0} \text{ m/s}. \quad (6.61)$$

Since $v = \omega r \Rightarrow \omega = v/r$ we thus have that the angular velocity of the pulley at time t_f has magnitude $\omega = \sqrt{5.0}$ rad/s.

Exercises

(i) Two uniform horizontal disks D_1 and D_2 have moments of inertia I_1 and I_2, respectively. They are rotating about a common axis z that goes through their centers and is perpendicular to their planes. The disks are a small vertical distance apart and are rotating in opposite directions with angular velocities of magnitude ω_1 and ω_2, respectively. By exerting an upward force on D_2, we move the disk up so that it comes in contact with D_1 and the two disks stick together. Find the system's kinetic energy before and after the disks stick together.

(ii) A uniform cylindrical drum of radius $R = 0.2$ m has a moment of inertia of 1.0 kgm^2 about an axis that passes through its center O and is perpendicular to the plane of the drum. A rigid, massless rod of length

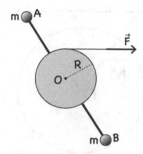

Figure 6.6 Top-down view of a system consisting of a drum, a rigid rod, and two small masses. At time $t_0 = 0$ a force $\vec{F}$ is exerted on a massless and inelastic rope wound around the drum and the whole system starts rotating.

$AB = 1.0$ m is attached to the drum with its center also at O, and with point masses of $m = 3.0$ kg firmly attached to each end. The system is resting on a table and is shown from above in Figure 6.6. At time $t_0 = 0$ a constant force $\vec{F}$ is applied to the drum by means of a massless, inelastic cord wrapped around its rim. At $t_1 = 10$ s, the system has an angular velocity of magnitude $\omega_1 = 4\pi$ rad/s. Ignore any frictional forces and find (a) the number of revolutions made by the system during the first 10 s of its motion, (b) the radial and tangential components of the linear acceleration of one of the 3.0 kg masses at time t_1, (c) the kinetic energy of the system at t_1, (d) the magnitude of the force $\vec{F}$, and (e) the power delivered to the system by $\vec{F}$ at time t_1.

(iii) A junction is used to attach one end of a uniform, thin rod of mass $M = 3.0$ kg and length $L = 1.0$ m to the ceiling. The rod is initially in equilibrium as it hangs vertically from the ceiling, but can rotate about a horizontal axis that passes through the end that connects it to the junction. At time $t_0 = 0$ we exert a force $\vec{F}$ at the rod's other (free) end that causes the rod to rotate. The force has a constant magnitude of $F = 30$ N and its direction always remains perpendicular to the length of the rod. At time t_1 the rod is horizontal and the magnitude of its angular velocity is $\omega = 2.0$ rad/s. Find the work of the frictional force exerted on the rod by the junction during the time interval between t_0 and t_1. You are given that the moment of inertia of a uniform rod about an axis going through one end is $I = \frac{1}{3}ML^2$ and that the gravitational acceleration is $g = 10$ m/s^2.

(iv) A uniform pulley of radius R and mass $M = 4.0$ kg is free to rotate (no friction) about a horizontal axis that goes through its center O and is perpendicular to the plane of the pulley. A massless and inelastic rope is wound around the pulley at its circumference and a mass $m = 2.0$ kg is

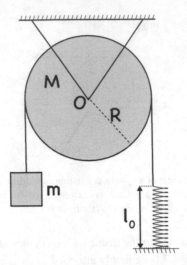

Figure 6.7 A massless, inelastic rope is wound around a uniform pulley. One end
of the rope is attached to a hanging mass m while the other is attached to a vertical
spring, which is initially at its natural length l_0. At time $t_0 = 0$ the system is
released from rest and the mass begins to fall, which causes the spring to stretch.

hanging from one free end of the rope, while the other is attached to a
spring anchored to the ground, as shown in Figure 6.7. The spring is ideal
and has a spring constant of $k = 100$ N/m. Initially, the spring is at its
natural length l_0 and the two ends of the rope are taut. At time $t_0 = 0$ we
release the system from rest. Find (a) by how much the spring has
stretched at the time t_1 when the hanging mass has its maximum velocity,
(b) the magnitude of this maximum velocity, and (c) the maximum
magnitude of the force exerted on the ground by the spring. Assume that
the rope does not slip on the rotating pulley. You are given the
gravitational acceleration $g = 10$ m/s^2 and that the pulley's moment of
inertia about its axis of rotation is $I = \frac{1}{2}MR^2$.

7

Combining Translation and Rotation

A First Look

So far in this book, we have looked at rigid bodies constrained to rotate about a fixed axis. In this chapter, we want to combine the techniques we have developed with what we already know from introductory physics about translational motion. The goal is to study objects which execute both translational and rotational motion simultaneously, a motion that at first glance might appear quite complicated. Examples of such a motion include a spinning tennis ball traveling through the air or a flying Frisbee.

7.1 General Motion of a Rigid Body: Chasles' Theorem

Let's consider a uniform rod one more time. Imagine that we throw the rod in such a way that it spins as it moves through the air and that we take a video of its motion. Snapshots of this video are shown schematically in Figure 7.1. Analysis of the video will reveal that the center of mass of the rod obeys the equations of motion for a point mass projectile, while the rest of the rod rotates about an axis that goes through the rod's center of mass as it travels through the air. Chasles' theorem tells us that we can study the general motion of any rigid body by combining two independent motions: One is the translational motion of the center of mass, and the second is the rotation of all points of the body about an axis through the center of mass. In this chapter, we will apply this idea to the special case of rolling motion, which we have seen in our introductory physics course. In upper level mechanics classes, you will prove Chasles' theorem and develop this idea further, but the simple case of rolling motion serves as a good introduction.

7.2 The Kinematics of Rolling Motion

Imagine we are observing a uniform wheel rolling to the right along a horizontal surface. Applying Chasles' theorem, we can see that the center of mass

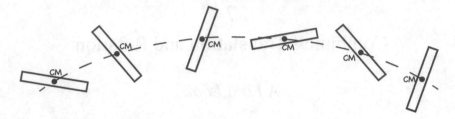

Figure 7.1 A uniform rod is thrown at an angle with respect to the horizontal in such a way that it spins as it moves through the air. According to Chasles' theorem, the motion of the rod is a combination of two independent motions: The translational motion of the center of mass, which in this case is that of a projectile, and the rotational motion of all other points of the rod about an axis that goes through its center of mass.

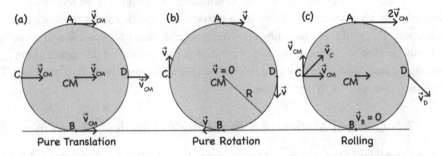

Figure 7.2 Rolling motion of a uniform wheel as a composition of two motions: Pure translation and pure rotation. (a) For a rigid body in pure translation, all points move with the same speed, the speed of the center of mass. (b) For pure rotation, each point of the rigid body has a linear velocity whose magnitude can be determined from its angular velocity via $v = \omega r$. (c) In rolling motion, the instantaneous velocity of any point is the vector sum of the velocity of the center of mass and the point's tangential velocity associated with its rotation about the center of mass. In the case of no slipping, Equation (7.4) gives that the tangential speed of the points on the rim due to the rotation is equal to the translational speed of the center of mass.

(CM) of the wheel is clearly executing translational motion (since it is moving to the right), and at the same time, all other points of the wheel are rotating about an axis that goes through the center of mass and is perpendicular to the plane of the wheel. In this way, the composition of translation (Figure 7.2a) and rotation (Figure 7.2b) gives rolling motion (Figure 7.2c). Using Figure 7.2 as our guide, we will now look at how to analyze rolling motion by examining its translational and rotational components.

Translation (Figure 7.2a): Since the wheel is rigid, all its points have the *same* translational velocity, the velocity of the center of mass $\vec{v}_{cm}$. If they did not, the object would deform (i.e., would not be rigid) since different parts would be moving to the right at different speeds. The displacement of any point on the wheel is the same as the displacement of the center of mass and is given simply by 1D kinematics.

Rotation (Figure 7.2b): All points of the rotating wheel have linear velocities whose direction is *tangent* to their circular trajectory and whose magnitude is proportional to the radius of their trajectory r via the equation $v = \omega r$. As discussed in Chapter 1, this implies that at the axis of rotation $v = 0$. It also implies that all points on the rim ($r = R$) have the same speed $v = \omega R$. The angular displacement of any point on the wheel is given simply by kinematics for rotational motion.

Rolling (Figure 7.2c): As the wheel rolls, the velocity of any point on the wheel is given by the vector sum of its translational velocity (i.e., the velocity of the center of mass) and its tangential velocity associated with the rotational motion of the point about the center of mass. Because the wheel is rigid, the relative positions between points of the wheel do not change. Thus, we can find the displacement of any point on the wheel if we know the position of the point relative to the center of mass, the displacement of the center of mass, and the number of rotations the wheel has undergone during the rolling motion.

In addition to considering rolling motion as a combination of translation of the center of mass and rotation about an axis through the center of mass, under certain conditions (no slipping) we can also think of it as *pure* rotation about an instantaneous axis that goes through the point of the rolling object in contact with the surface (point B in Figure 7.2). This is because if the wheel does not slip, the instantaneous velocity of point B is zero (shown in Figure 7.2c), as we will discuss in Section 7.2.1, and thus all other points of the wheel instantaneously rotate about that point. Both interpretations will help us develop an intuition for and understanding of rolling motion.

We will now examine what happens when the wheel is rolling without slipping. The case of rolling with slipping will be briefly discussed in Section 7.2.2.

7.2.1 Rolling without Slipping

Imagine that the uniform wheel shown in Figure 7.3 rolls to the right on a horizontal surface. At time t point A is in contact with the ground and at time $t + \Delta t$ point B is in contact with the ground. If the wheel rolls without slipping, then as it rotates by an arc length $AB = \Delta S$ during the time interval Δt, the center of mass will move to the right by an amount Δx_{cm}, where

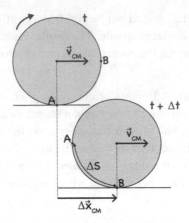

Figure 7.3 If a uniform wheel is rolling without slipping, then as the wheel rotates by an arc length ΔS during an interval Δt, its center of mass will move to the right by an amount $\Delta S = \Delta x_{cm}$. Though the wheel is rolling on a horizontal surface, the snapshot at time $t + \Delta t$ is vertically offset from the snapshot at time t to show the one-to-one correspondence between the arc-length points and the horizontal surface points which the wheel contacts.

$$\Delta S = \Delta x_{cm}. \tag{7.1}$$

If we take the limit as Δt goes to zero, the arc length ΔS becomes dS, the displacement Δx_{cm} becomes dx_{cm}, and Equation (7.1) becomes:

$$dS = dx_{cm}. \tag{7.2}$$

The physical explanation for Equations (7.1) and (7.2) is as follows: As the wheel rolls without slipping, all points of the wheel's rim within the arc length $AB = \Delta S$ will consecutively make contact with the horizontal surface; that is, there will be a one-to-one correspondence between the arc length points and the horizontal surface points which the wheel contacts. Another way of thinking about this is by imagining that the rim is painted. As the wheel rolls, the part of the wheel that comes in contact with the surface will transfer the paint onto the horizontal surface. At the end, the length of the paint streak on the ground will be equal to the arc length that has lost its paint.

Equation (7.2) is an important equation because it quantifies the constraint of rolling without slipping. If this equation were not true, then none of the equations we will derive in this section would be true. Equation (7.2) does not apply if there is slipping because, in this case, the wheel will rotate some without moving to the right (or it will move to the right without rotating) so that there is not a one-to-one correspondence between the points along the arc length and the points with which they make contact on the horizontal surface. Thus, the arc length associated with the wheel's rotation will not equal the displacement of the center of mass along the surface.

If we divide both sides of equation (7.2) by the infinitesimal time dt that it took for the wheel to rotate by this amount dS, we will have:

$$\frac{dS}{dt} = \frac{dx_{cm}}{dt} \Rightarrow v_{\text{points on rim}} = v_{cm}. \qquad (7.3)$$

But, as we discussed, $v_{\text{points on rim}} = \omega R$. So,

$$v_{cm} = \omega R = v_{\text{points on rim}}. \qquad (7.4)$$

Equation (7.4) connects the translational motion of the center of mass of the wheel to the rotational motion of the points on the rim. If we take the derivative with respect to time we get:

$$\frac{dv_{cm}}{dt} = \frac{d(\omega R)}{dt}$$

$$\Rightarrow a_{cm} = R\frac{d\omega}{dt} = \alpha R. \qquad (7.5)$$

Again, we keep in mind that these equations are true only if there is no slipping.

Now let's revisit Figure 7.2c, which shows a wheel executing rolling motion without slipping, and first focus on point B. This point, like all points of the wheel, has a velocity associated with each of its motions: In its translational motion, point B will move linearly with the wheel's velocity $\vec{v}_{cm}$; in its rotational motion, it will rotate with a tangential velocity of $v = \omega R$. At point B these two velocities have opposite directions and, because of Equation (7.4), they have the same magnitude. Therefore, the total velocity at point B is $v_B = 0$. If we now look at point A, its translational velocity has the same direction as the tangential velocity due to its rotational motion and, because of Equation (7.4), we see that $\vec{v}_A = 2\vec{v}_{cm}$. In a similar way we can show that both $\vec{v}_C$ and $\vec{v}_D$ form angles of 45° with the horizontal and have magnitudes $v_C = v_D = v_{cm}\sqrt{2}$. For points on the wheel that are not on the rim (i.e., $r < R$), the rotational speed is $v = \omega r$ and is less than their translational speed $v_{cm} = \omega R$.

Now let's examine the uniform rolling wheel in Figure 7.4 to study its acceleration as a combination of pure translation of the center of mass and pure rotation about an axis through the center of mass.

Translation: Since the wheel is a rigid body, all points of the wheel have the *same* translational acceleration, the acceleration of the center of mass $\vec{a}_{cm} = \frac{d\vec{v}_{cm}}{dt}$. If they did not, then the wheel would deform and, thus, could not be considered a rigid body.

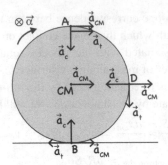

Figure 7.4 The total acceleration of a point on a uniform wheel that is rolling without slipping is the vector sum of the components associated with the translation of the center of mass ($\vec{a}_{cm}$) and the rotation of the points of the wheel about an axis through the center of mass ($\vec{a}_c$ and $\vec{a}_t$).

Rotation: As discussed in Chapter 1, all points of the wheel have two commonly used components of the acceleration associated with rotation: The tangential acceleration $\vec{a}_t$ and the centripetal acceleration $\vec{a}_c$. The tangential component shows how quickly the speed of rotation, $v = \omega r$, changes (i.e., how fast the wheel speeds up or slows down). By taking the derivative of this equation with respect to time, we get:

$$\frac{dv}{dt} = \frac{d(\omega r)}{dt}$$

$$\Rightarrow a_t = r\frac{d\omega}{dt} = \alpha r, \tag{7.6}$$

where we have used the definition of angular acceleration and the fact that r remains constant over time for a given point of the wheel. For points on the rim, where $r = R$, $a_t = \alpha R$. However, as we just derived in Equation (7.5), $a_{cm} = \alpha R$. So, *for points on the rim, when there is no slipping*:

$$a_t = a_{cm} = \alpha R. \tag{7.7}$$

On the other hand, the centripetal component is responsible for keeping the points on the wheel moving in a circle; in other words, it causes the *direction* of $\vec{v}$ to change. For the centripetal acceleration, we know that

$$a_c = \frac{v^2}{r} = \frac{\omega^2 r^2}{r} = \omega^2 r, \tag{7.8}$$

where we have used $v = \omega r$.

To find the total linear acceleration of any point on the rotating wheel, we combine the components of acceleration from translation and rotation:

$$\vec{a}_{\text{total}} = \underbrace{\vec{a}_{cm}}_{\text{for translation}} + \underbrace{\vec{a}_t + \vec{a}_c}_{\text{for rotation}}. \tag{7.9}$$

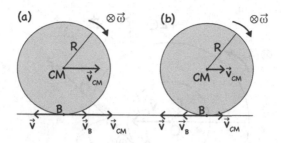

Figure 7.5 If a uniform wheel is slipping while rolling clockwise, then it can slip both forward (a) and backward (b). In (a), the net velocity at point B will be pointing forward (to the right) while in (b) it will be pointing backward (to the left).

7.2.2 What Happens if the Wheel Is Slipping While Rolling?

Figure 7.5 shows a uniform wheel of radius R which is *slipping* while rolling on a horizontal surface. At time t the velocity of its center of mass is $\vec{v}_{cm}$ and the magnitude of the linear velocity for a point on the rim due to rotational motion is $v = \omega R$. Because the wheel is slipping, Equation (7.2) is no longer true. Consequently, none of the subsequent equation in Section 7.2.1 are true and so $v_{cm} \neq \omega R$. Thus, the point of the wheel that is in contact with the horizontal surface (point B of Figure 7.5) now has a total velocity that is *not* zero ($v_B \neq 0$). Let's look at two possible cases:

(a) If the wheel is rolling clockwise and slipping forward, (Figure 7.5a), the net velocity at B ($\vec{v}_B$) is pointing forward and has magnitude:

$$\vec{v}_B = \vec{v}_{cm} + \vec{v}$$

$$\Rightarrow v_B = v_{cm} - v = v_{cm} - \omega R. \tag{7.10}$$

(b) If the wheel is rolling clockwise and slipping backward (Figure 7.5b), the net velocity at B ($\vec{v}_B$) is pointing backward and has magnitude:

$$\vec{v}_B = \vec{v}_{cm} + \vec{v}$$

$$\Rightarrow v_B = |v_{cm} - v| = \omega R - v_{cm}. \tag{7.11}$$

7.2.3 Kinematics of a Spool of Thread

In Figure 7.6 we have a spool of thread which we model as a uniform disk of mass M and radius R. A massless, inelastic rope is wrapped around the disk at a groove a distance r from the disk's center O. Let's study the disk as it rolls without slipping on a rough, horizontal surface due to a force $\vec{F}$ exerted at the free end A of the rope. $\vec{F}$ forms an angle ϕ with the horizontal. Since the rope is

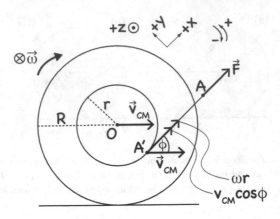

Figure 7.6 A spool of thread modeled as a uniform disk of radius R and mass M that can roll on a rough, horizontal surface. The disk has a massless, inelastic rope wrapped around it at a groove a distance r from its center O. We exert a force $\vec{F}$ at the free end A of the rope at an angle ϕ relative to the horizontal.

inelastic, this force is effectively transferred and exerted at point A' and causes the rope to unwind without slipping against the groove. The angle ϕ determines the direction of motion of the disk, which we will discuss in more detail when we talk about the dynamics of the disk's motion in Section 7.3.2. For now, let's assume it is moving to the right and rotating in the clockwise direction and examine the relationship between the length of the rope that is unwound and the distance the disk has rolled.

At time t, the disk has angular velocity $\vec{\omega}$, angular acceleration $\vec{\alpha}$, and its center of mass has linear velocity $\vec{v}_{cm}$ and linear acceleration $\vec{a}_{cm}$. Also, due to rotation, all points of the groove have tangential velocity $v = \omega r$, all point on the rim have tangential velocity $v = \omega R$, and since the disk does not slip on the surface, we know from Equation (7.4) that $v_{cm} = \omega R$.

Let's define the x axis to be along the rope (not along the horizontal). Since the rope is inelastic and does not slip, the total linear velocity (i.e., from both translation and rotation) along the x axis of any point on the rope (e.g., point A) is equal to the total linear velocity of point A' along the x axis:

$$\vec{v}_A = \vec{v}_{A'} = \underbrace{\vec{v}_{cm_x}}_{\text{translational}} + \underbrace{\vec{v}_{\text{rot}}}_{\text{rotational}}$$

$$\Rightarrow v_A = v_{A'} = v_{cm} \cos \phi + \omega r$$

$$= v_{cm} \cos \phi + \frac{v_{cm}}{R} r$$

$$= v_{cm} \left(\cos \phi + \frac{r}{R} \right). \tag{7.12}$$

Based on this result, we can find the total linear acceleration of any point on the rope and, thus, of point A by taking the time derivative of its velocity:

$$a_A = \frac{dv_A}{dt} = \frac{dv_{cm}}{dt}\left(\cos\phi + \frac{r}{R}\right) = a_{cm}\left(\cos\phi + \frac{r}{R}\right). \qquad (7.13)$$

The displacement of point A in the direction of the rope (i.e., along the x axis) dx_A during time dt can be found by multiplying both sides of our result for v_A in Equation (7.12) with dt:

$$dx_A = v_A dt = v_{cm}dt\left(\cos\phi + \frac{r}{R}\right) = dS_{cm}\left(\cos\phi + \frac{r}{R}\right), \qquad (7.14)$$

where $dS_{cm} = v_{cm}dt$ is the translational displacement of the disk (i.e., the displacement along the ground). The length of the rope dl that gets unwrapped during time dt equals the arc length traversed by point A' on the groove:

$$dl = rd\theta = r\frac{dS_{cm}}{R}, \qquad (7.15)$$

where we have used the fact that $d\theta$ can be expressed in terms of either the radius r of the groove or the radius R of the disk (i.e., $d\theta = \frac{dl}{r} = \frac{dS}{R}$) and that there is no slipping of the disk against the surface (i.e., $\frac{dS}{R} = \frac{dS_{cm}}{R}$).

The last two results show that $dx_A \neq dl$. In other words, the amount of rope that has been unwound (dl) is not the same as the amount by which the end of the rope has moved along the x axis (dx_A). Why is that? The reason is that as the rope gets unwound, the disk is also moving to the right. Notice from Equations (7.14) and (7.15) that $dx_A = dS_{cm}\cos\phi + dl$. Therefore, point A moves the length of the unwound rope ($r\frac{dS_{cm}}{R}$) plus a distance equal to the component along the rope of the translational displacement dS_{cm} of the disk ($dS_{cm}\cos\phi$).

7.3 The Dynamics of Rolling Motion

In Chapter 2 we extensively studied the condition for rotational equilibrium which, together with translational equilibrium, allowed us to study objects that are either stationary or moving with constant velocity. In Chapter 3 we examined Newton's Second Law for rotational motion by applying what we know from Newton's Second Law for translational motion. Since rolling motion is the combination of translational and rotational motion, then we can have the following cases depending on whether we have translational and/or rotational equilibrium and on whether the net force and/or net torque on the object are constant:

- $\Sigma\vec{F} = 0$ and $\Sigma\vec{\tau} = 0$: The object is in both translational and rotational equilibrium. This implies that $\vec{v}_{cm} = $ constant (or $\vec{v}_{cm} = 0$) and $\vec{\omega} = $ constant (or $\vec{\omega} = 0$). If both $\vec{v}_{cm} = 0$ and $\vec{\omega} = 0$ then the object is stationary.

- $\Sigma\vec{F}$ = constant $\neq 0$ and $\Sigma\vec{\tau} = 0$: The translational motion is $1D$ motion with constant acceleration $\vec{a}_{cm} = \Sigma\vec{F}/m$. However, there is rotational equilibrium, so $\vec{\omega}$ = constant (or $\vec{\omega} = 0$).
- $\Sigma\vec{F} = 0$ and $\Sigma\vec{\tau}$ = constant $\neq 0$: Because there is no net force, there is translational equilibrium, so $\vec{v}_{cm}$ = constant (or $\vec{v}_{cm} = 0$). The rotational motion has constant angular acceleration $\vec{\alpha} = \Sigma\vec{\tau}/I$.
- $\Sigma\vec{F}$ = constant $\neq 0$ and $\Sigma\vec{\tau}$ = constant $\neq 0$: The translational motion is $1D$ motion with constant acceleration $\vec{a}_{cm} = \Sigma\vec{F}/m$. The rotational motion has constant angular acceleration $\vec{\alpha} = \Sigma\vec{\tau}/I$.
- $\Sigma\vec{F} \neq$ constant and $\Sigma\vec{\tau} \neq$ constant: For the translational motion $\vec{a}_{cm} \neq$ constant and so the expression $\vec{a}_{cm} = \Sigma\vec{F}/m$ can only give the *instantaneous* acceleration. Thus, the equations for $1D$ motion with constant acceleration do not apply. Similarly, for the rotational motion $\vec{\alpha} \neq$ constant and so the expression $\vec{\alpha} = \Sigma\vec{\tau}/I$ can only give the *instantaneous* angular acceleration. Thus, the equations for rotational motion with constant angular acceleration do not apply.

Now that we have an idea of how Newton's First and Second Laws apply to rolling motion, we can study how various forces affect the dynamics of this motion.

7.3.1 Direction of the Static Friction in Rolling Motion

Suppose a uniform disk is rolling without slipping and accelerating ($v_{cm_2} > v_{cm_1}$) down an inclined plane, as shown in Figure 7.7a. The rotational motion about the axis through the center of mass must be due to the torque provided by the static friction $\vec{f}_s$. The only other forces, $\vec{F}_g$ and $\vec{n}$, have lines of action through the axis of rotation and thus cannot provide torque about this axis. There is no kinetic friction since the disk is not sliding against the incline.

Since the only force responsible for the disk's rotation is $\vec{f}_s$, the direction of $\vec{f}_s$ depends on the direction of the change in angular velocity $\Delta\vec{\omega}$ of the disk. Let's think about why this is true: Since $\vec{\alpha} = \frac{\Delta\vec{\omega}}{\Delta t}$, $\vec{\alpha}$ and $\Delta\vec{\omega}$ have the same direction. Since $\Sigma\vec{\tau} = I\vec{\alpha}$ and $I > 0$ then $\Sigma\vec{\tau}$ has the same direction as $\vec{\alpha}$ and, thus, the same direction as $\Delta\vec{\omega}$. Since $\Sigma\vec{\tau} = \vec{\tau}_{fs}$, this means that $\vec{\tau}_{fs}$ has the same direction as $\Delta\vec{\omega}$. The direction of the position vector $\vec{r}$ for $\vec{f}_s$ is fixed and points from the axis of rotation to the point of contact of the disk with the incline. Therefore, there is only one direction for $\vec{f}_s$ that will give, via the right-hand rule, a torque $\vec{\tau}_{fs}$ that has the same direction as $\Delta\vec{\omega}$. If the disk is rolling clockwise and is accelerating down the incline (because of the component of $\vec{F}_g$ along the incline), then $\vec{\omega}$ points into the page and increases

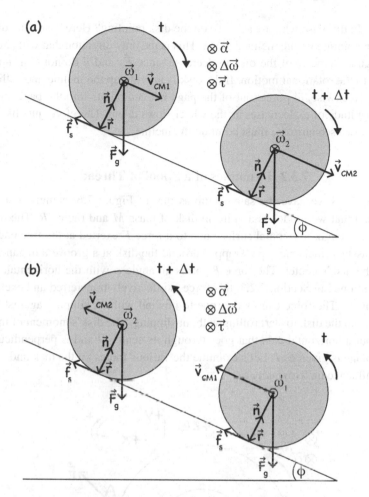

Figure 7.7 A uniform disk is rolling without slipping (a) down and (b) up an inclined plane. We can decompose the motion into the translation of the center of mass and the rotation about an axis through the center of mass. The rotational motion is due to the torque provided by the static friction, since all other forces have lines of action through the axis of rotation. Therefore, the direction of the static friction must be such that the direction of its torque is the same as that of the disk's change in angular velocity.

in magnitude. Thus, $\Delta\vec{\omega}$ also points into the page. This implies that $\vec{f}_s$ must point up the incline, as shown.

For another scenario, let's now look at the disk as it is rolling without slipping up the inclined plane, as shown in Figure 7.7b. The disk slows down in the process ($v_{cm_2} < v_{cm_1}$), because of the component of $\vec{F}_g$ that points down the incline. In this case which way is $\vec{f}_s$ on the disk pointing? Here $\vec{f}_s$ on the disk is again pointing up the inclined plane. This is the only direction that will cause the angular velocity of the disk to decrease since $\vec{F}_g$ and $\vec{n}$ do not contribute torque to the rotational motion. If the disk is moving up the incline and rolling counterclockwise, $\vec{\omega}$ points out of the page but $\Delta\vec{\omega}$ points into the page since the magnitude of $\vec{\omega}$ decreases as the wheel slows down. Therefore, just like in the previous scenario, $\vec{f}_s$ must point up the incline.

7.3.2 Dynamics of a Spool of Thread

In Figure 7.8 we have the same setup as that of Figure 7.6, namely a spool of thread that we model as a uniform disk of mass M and radius R. The disk rolls on a rough, horizontal surface due to a force $\vec{F}$ exerted at the free end A of a massless, inelastic rope wrapped around the disk at a groove a distance r from the disk's center. The force $\vec{F}$ forms an angle ϕ with the horizontal. As we mentioned in Section 7.2.3, this force is effectively transferred and exerted at point A'. The force causes the rope to unwind without slipping against the groove and the disk to start rolling without slipping. The disk's moment of inertia about a horizontal axis that goes through its center O and is perpendicular to its plane is given as I. Let's examine the various forces on the disk and how they influence its dynamics.

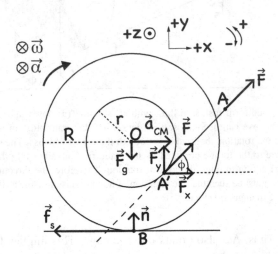

Figure 7.8 A disk of radius R that can roll on a rough, horizontal surface has a massless, inelastic rope wrapped around it at a groove a distance r from the disk's center O. We exert a force $\vec{F}$ at the free end A of the rope at an angle ϕ relative to the horizontal. For the disk to rotate clockwise about an axis going through O the static friction must point to the left.

For starters, we again assume that the disk is moving to the right and rotating in the clockwise direction, as shown in Figure 7.8. In contrast to our choice of coordinate system in Figure 7.6, we now pick our $+x$ axis to point to the right along the disk's direction of translational motion.

But how can we justify this assumption about the direction of motion being to the right? For a moment, let's think about the rotation of all points of the disk about the instantaneous axis of rotation that is perpendicular to the plane of the page and goes through point B (i.e., the point in contact with the ground). By the right-hand rule, the torque due to $\vec{F}$ about this axis points into the page, which means the angular acceleration $\vec{\alpha}$ points into the page and so the disk rotates clockwise. Since the disk does not slip, this means that the disk accelerates to the right. So, our assumption is justified. Now we can see how the angle ϕ determines the direction of motion of the disk: If ϕ increases to a value such that the line of action of the force $\vec{F}$ intersects the horizontal ground at a point to the right of point B, then by the right-hand rule, the torque due to $\vec{F}$ about this axis through B points out of the page, which means the angular acceleration $\vec{\alpha}$ points out of the page and so the disk rotates counterclockwise. Since the disk does not slip, this means that the disk accelerates to the left.

Now, let's return to considering the motion of the disk as a combination of translational motion of the center of mass and rotation about the center of mass. If we apply the right-hand rule to find the torque due to $\vec{F}$ about the axis through the center of the disk O, we see that $\vec{\tau}_F$ points out of the page (i.e., in the $+z$ direction) and therefore causes counterclockwise rotation. Since the disk is rotating clockwise, it must be that $\vec{f}_s$ provides a torque about this axis of rotation that points into the page. Thus, $\vec{f}_s$ must point to the left, as shown in Figure 7.8.

At time t, the disk has angular velocity $\vec{\omega}$, angular acceleration $\vec{\alpha}$, and its center of mass has linear velocity $\vec{v}_{cm}$ and linear acceleration $\vec{a}_{cm}$. Also, all points of the groove have tangential acceleration $a = \alpha r$, all point on the rim have tangential acceleration $a = \alpha R$, and since the disk does not slip on the surface, we know from Equation (7.5) that $a_{cm} = \alpha R$. We follow the now familiar methodology and look at each type of motion separately.

Translation:

y axis: This disk is in equilibrium along this axis. Therefore,

$$\Sigma \vec{F}_y = 0$$
$$\Rightarrow n + F_y - F_g = 0$$
$$\Rightarrow n + F \sin \phi = Mg. \tag{7.16}$$

x axis: The disk is accelerating along the x axis, so we can apply Newton's Second Law:

$$\Sigma \vec{F}_x = M\vec{a}_{cm}$$

$$\Rightarrow F_x - f_s = Ma_{cm}$$

$$\Rightarrow F\cos\phi - f_s = Ma_{cm}$$

$$\Rightarrow f_s = F\cos\phi - Ma_{cm}. \tag{7.17}$$

Note that we cannot say here that $f_s = \mu_s n$. That would correspond to the maximum static friction and we cannot be sure at this point that the magnitude of the static friction is maximum.

Rotation:

We can apply Newton's Second Law for rotational motion:

$$\Sigma \vec{\tau}_O = I\vec{\alpha}$$

$$\Rightarrow \underbrace{\tau_{F_g}}_{0\,(r=0)} + \underbrace{\tau_n}_{0\,(\phi=\pi)} -\tau_{f_s} + \tau_F = -I\alpha$$

$$\Rightarrow -f_s R \sin\left(\frac{\pi}{2}\right) + Fr \sin\left(\frac{\pi}{2}\right) = -I\alpha$$

$$\Rightarrow f_s R - Fr = I\alpha. \tag{7.18}$$

Using $a_{cm} = \alpha R$ and substituting f_s from Equation (7.17) into Equation (7.18) gives:

$$f_s R - Fr = I\frac{a_{cm}}{R}$$

$$\Rightarrow (F\cos\phi - Ma_{cm})R - Fr = I\frac{a_{cm}}{R}. \tag{7.19}$$

Lastly, we solve for a_{cm}:

$$FR\cos\phi - Fr = I\frac{a_{cm}}{R} + Ma_{cm}R$$

$$\Rightarrow FR\left(\cos\phi - \frac{r}{R}\right) = I\frac{a_{cm}}{R}\left(1 + \frac{MR^2}{I}\right)$$

$$\Rightarrow a_{cm} = \frac{FR^2\left(\cos\phi - \frac{r}{R}\right)}{I\left(1 + \frac{MR^2}{I}\right)}. \tag{7.20}$$

This final result confirms what we discussed qualitatively earlier, namely that the direction of the acceleration depends on the value of ϕ. For a constant ratio of $\frac{r}{R}$, if ϕ increases such that $\cos\phi < \frac{r}{R}$, the acceleration comes out to be negative. Since the x component of $\vec{F}$ is to the right, if the acceleration a_{cm} is to the left then the static friction must again point to the left to provide this leftward acceleration. Now let's do an example to conclude our discussion of kinematics and dynamics in rolling motion.

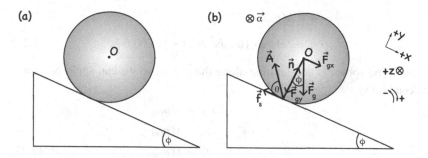

Figure 7.9 (a) A uniform solid sphere is initially at rest on top of an inclined plane of angle $\phi = 30°$. (b) The sphere rolls without slipping down the incline. The static friction $\vec{f}_s$ must point upward to provide the necessary torque for the sphere to rotate clockwise about an axis through O.

Example 1: Sphere rolling down a ramp. A uniform solid sphere of mass $M = 7.0\,\text{kg}$ and radius R is initially at rest on top of an inclined plane of angle $\phi = 30°$, as shown in Figure 7.9a. At time $t_0 = 0$ it begins to roll without slipping down the incline. Find (a) the acceleration of the sphere's center of mass, (b) the vertical distance h that the sphere travels between time $t_0 = 0$ and time $t = 1.4$ s, (c) the reaction force $\vec{A}$ exerted on the sphere by the inclined plane, and (d) the minimum coefficient of static friction μ_s between the sphere and the incline. You are given that $g = 10\,\text{m/s}^2$ and the moment of inertia about an axis through the sphere's center of mass is $I_{cm} = \frac{2}{5}MR^2$.

As always, we start by drawing the forces and the coordinate system (Figure 7.9b). The forces exerted on the sphere are the gravitational force $\vec{F}_g$, the reaction force $\vec{n}$ from the inclined plane, and the static friction $\vec{f}_s$ from the plane on the sphere. Based on our previous discussion, we can see why $\vec{f}_s$ points upward along the incline – it provides the torque for the sphere to rotate clockwise about an axis through O. Note that the total reaction force $\vec{A}$ is the vector sum of $\vec{n}$ and $\vec{f}_s$:

$$\vec{A} = \vec{n} + \vec{f}_s. \tag{7.21}$$

To study the motion of the sphere, we again decompose it into translation and rotation.

Translation:

y axis: The sphere is in equilibrium along this axis. Therefore,

$$\Sigma \vec{F}_y = 0$$

$$\Rightarrow n - F_{g_y} = 0$$

$$\Rightarrow n = Mg \cos \phi$$

$$\Rightarrow n = 7.0 \text{ kg} \cdot 10 \text{ m/s}^2 \cos\left(\frac{\pi}{6}\right) = 60.6 \text{ N}. \tag{7.22}$$

x axis: The sphere is accelerating along the *x* axis, so we can apply Newton's Second Law:

$$\Sigma \vec{F}_x = M \vec{a}_{cm}$$

$$\Rightarrow F_{gx} - f_s = M a_{cm}$$

$$\Rightarrow Mg \sin \phi - f_s = M a_{cm}. \tag{7.23}$$

Again we cannot say here that $f_s = \mu_s n$. That would correspond to the maximum static friction and we cannot be sure at this point that the magnitude of the static friction is maximum.

Rotation:

We can apply Newton's Second Law for rotational motion:

$$\Sigma \vec{\tau}_O = I \vec{\alpha}$$

$$\Rightarrow \underbrace{\tau_{F_g}}_{0} + \underbrace{\tau_n}_{0} + \tau_{f_s} = I \alpha$$

$$\Rightarrow f_s R \sin\left(\frac{\pi}{2}\right) = \frac{2}{5} M R^2 \alpha$$

$$\Rightarrow f_s = \frac{2}{5} M R \alpha. \tag{7.24}$$

Here we have used that the torque due to $\vec{F}_g$ is zero because it is applied at the axis of rotation, and that the torque due to $\vec{n}$ is zero because its line of action passes through the axis of rotation. Since the sphere rolls without slipping, $a_{cm} = \alpha R$, and Equation (7.24) becomes:

$$f_s = \frac{2}{5} M a_{cm}. \tag{7.25}$$

(a) If we plug Equation (7.25) into Equation (7.23) we can solve for a_{cm}:

$$Mg \sin \phi - \frac{2}{5} M a_{cm} = M a_{cm}$$

$$\Rightarrow g \sin \phi = \frac{7}{5} a_{cm}$$

$$\Rightarrow a_{cm} = \frac{5}{7} g \sin \phi$$

$$\Rightarrow a_{cm} = \frac{5}{7} \cdot 10 \text{ m/s}^2 \sin\left(\frac{\pi}{6}\right) = \frac{25}{7} \text{ m/s}^2. \tag{7.26}$$

(b) Since $\Sigma \vec{F}_x$ is constant, then $\vec{a} = \vec{a}_{cm}$ is constant. So, on the *x* axis we have 1*D* motion with constant acceleration, which is described by the equations:

$$\vec{v} = \underbrace{\vec{v}_0}_{0} + \vec{a}\Delta t \Rightarrow v = a\Delta t, \tag{7.27}$$

$$\Delta \vec{x} = \underbrace{\vec{v}_0}_{0} \Delta t + \frac{1}{2}\vec{a}\Delta t^2 \Rightarrow \Delta x = \frac{1}{2}a\Delta t^2. \tag{7.28}$$

Plugging the acceleration from part (a) into the equation for Δx gives:

$$\Delta x = \frac{1}{2} \cdot \frac{25}{7} \text{ m/s}^2 \cdot (1.4 \text{ s})^2 = \frac{7}{2} m. \tag{7.29}$$

Therefore,

$$h = \Delta x \sin \phi = \frac{7}{2} \text{ m} \cdot \sin\left(\frac{\pi}{6}\right) = \frac{7}{4} \text{ m}. \tag{7.30}$$

(c) We can also plug the acceleration from part (a) into Equation (7.25):

$$f_s = \frac{2}{5}Ma_{cm} = \frac{2}{5} \cdot 7.0 \text{ kg} \cdot \frac{25}{7} \text{ m/s}^2 = 10 \text{ N}. \tag{7.31}$$

Since we know $f_s = 10$ N and $n = 60.6$ N, and that $\vec{f}_s$ is perpendicular to $\vec{n}$, we can find the magnitude and direction of $\vec{A}$ as follows:

$$A = \sqrt{f_s^2 + n^2} = \sqrt{(60.6 \text{ N})^2 + (10 \text{ N})^2} = 61.4 \text{ N}, \tag{7.32}$$

and

$$\tan \theta = \frac{n}{f_s} = \frac{60.6 \text{ N}}{10 \text{ N}} \Rightarrow \theta = 1.4 \text{ rad}. \tag{7.33}$$

(d) We know that

$$f_s \leq f_{s,max} \Rightarrow f_s \leq \mu_s n \Rightarrow \mu_s \geq \frac{f_s}{n}. \tag{7.34}$$

Thus, at the minimum value of μ_s, the two sides will be equal:

$$\mu_{s,min} = \frac{f_s}{n} = \frac{10 \text{ N}}{60.6 \text{ N}} = 0.16. \tag{7.35}$$

7.4 Energy of Objects Executing both Translational and Rotational Motion

So far in this chapter, we have studied objects which translate and rotate simultaneously by considering these two motions independently. We will take the same approach when studying energy and its conservation in systems with this composite motion. As we know from introductory physics, objects moving linearly have a kinetic energy of $KE = \frac{1}{2}mv^2$. As we saw in Chapter 6, objects executing pure rotation have rotational kinetic energy of $KE_{rot} = \frac{1}{2}I\omega^2$. Considering the general motion of a rigid body as a combination of translation of the center of mass and rotation about an axis through the center of mass yields the total kinetic energy:

$$KE_{tot} = KE_{cm} + KE_{rot} = \frac{1}{2}mv_{cm}^2 + \frac{1}{2}I\omega^2. \qquad (7.36)$$

Let's now extend our definitions for the Work-Kinetic Energy (WKET), Conservation of Energy (CE), and Conservation of Mechanical Energy (CME) theorems to account for the kinetic energy of both translation and rotation and the work due to both a net force and a net torque. Starting from WKET, and including both translational and rotational terms, we have:

$$KE_{cm.i} + KE_{rot.i} + W_{net} + W_{net.torque} = KE_{cm.f} + KE_{rot.f}. \qquad (7.37)$$

If we replace the work associated with conservative forces (CF) and torques due to CF by the potential energy, we obtain CE:

$$KE_{cm.i} + KE_{rot.i} + U_i + W_{NCF} + W_{\tau_{NCF}} = KE_{cm.f} + KE_{rot.f} + U_f. \qquad (7.38)$$

In the case where there is no work due to nonconservative forces (NCF) and no work due to the torques of NCF, CE becomes CME:

$$KE_{cm.i} + KE_{rot.i} + U_i = KE_{cm.f} + KE_{rot.f} + U_f. \qquad (7.39)$$

Let's look at one interesting feature of the relationship between the kinetic energies in rolling motion before we finish the chapter with an example.

7.4.1 Comparison between Translational and Rotational Kinetic Energy of Rolling Objects

Imagine an object of mass M and radius R that is rolling without slipping on a rough surface. We will again consider this motion as a combination of the translation of the center of mass and a rotation of the object about an axis through its center of mass. Let's calculate the ratio of the translational kinetic energy to the rotational kinetic energy and examine the contribution of each to the total kinetic energy of the object.

Given that $KE_{cm} = \frac{1}{2}Mv_{cm}^2$, $KE_{rot} = \frac{1}{2}I\omega^2$, and $v_{cm} = \omega R$, we have for the ratio of the two energies:

$$\frac{KE_{cm}}{KE_{rot}} = \frac{\frac{1}{2}Mv_{cm}^2}{\frac{1}{2}I\omega^2} = \frac{M\omega^2 R^2}{I\omega^2} = \frac{MR^2}{I}. \qquad (7.40)$$

This result shows that the ratio of the two kinetic energies depends only on parameters involving the object's geometry and axis of rotation – the ratio is independent of the object's translational and rotational velocities. Let's consider some examples:

(i) For a uniform cylinder whose axis of rotation is the long axis of the cylinder, we know that $I_{cm} = \frac{1}{2}MR^2$. Then

$$\frac{KE_{cm}}{KE_{rot}} = \frac{MR^2}{\frac{1}{2}MR^2} = 2. \tag{7.41}$$

(ii) For a uniform ring whose axis of rotation is perpendicular to the plane of the ring, we know that $I_{cm} = MR^2$. Then

$$\frac{KE_{cm}}{KE_{rot}} = \frac{MR^2}{MR^2} = 1. \tag{7.42}$$

(iii) For a uniform solid sphere we know that $I_{cm} = \frac{2}{5}MR^2$. Then

$$\frac{KE_{cm}}{KE_{rot}} = \frac{MR^2}{\frac{2}{5}MR^2} = \frac{5}{2}. \tag{7.43}$$

(iv) For a uniform spherical shell we know that $I_{cm} = \frac{2}{3}MR^2$. Then

$$\frac{KE_{cm}}{KE_{rot}} = \frac{MR^2}{\frac{2}{3}MR^2} = \frac{3}{2}. \tag{7.44}$$

But what is the significance of these ratios? They tell us what percentage of the work done on a rolling object results in a change in the translational kinetic energy vs. the rotational kinetic energy. For example, if we do work to cause a spherical shell to roll without slipping, 60% of the energy transferred to the shell will become translational kinetic energy, and 40% will become rotational kinetic energy. In other words, the ratio between the two energies will be 3/2. Let's finish this chapter with an example.

Example 2: Rolling cylinder. Figure 7.10a shows a uniform cylinder of mass $M = 0.4$ kg and radius R. The cylinder is initially at rest on a horizontal surface with a massless and inelastic string wound around its circumference. At time $t_0 = 0$ we start exerting a constant horizontal force of magnitude $F = 10$ N on the string's free end A. The cylinder begins to roll on the horizontal surface without slipping. Find the magnitude of the velocity of the cylinder's center of mass at time t_1 when the cylinder's displacement from its initial position is $\Delta x_{cm} = 1.5$ m. The moment of inertia of the cylinder about its long axis (through point O and perpendicular to the plane of the page) is $I = \frac{1}{2}MR^2$.

Figure 7.10b shows our coordinate system and all the forces ($\vec{F}_g$, $\vec{n}$, $\vec{F}_T$, and $\vec{f}_s$) exerted on the cylinder. As we pull the string to the right with force $\vec{F}$, this

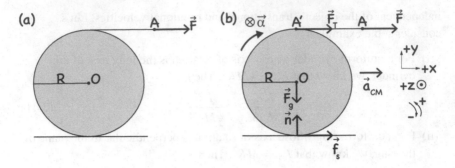

Figure 7.10 (a) A uniform cylinder has a massless and inelastic string wrapped around its circumference. We exert a force on the string's free end A and the cylinder begins to roll without slipping. (b) As the cylinder rolls, the force of static friction $(\vec{f_s})$ points in the same direction as the force of tension $(\vec{F_T})$ on the cylinder.

force is exerted on the cylinder at point A' via the force due to the tension $\vec{F_T}$ in the inelastic string. Although perhaps counter-intuitive, the static friction force points to the right in this case. Proving this is an exercise left for the reader at the end of the chapter.

Since the rolling of the cylinder can be decomposed into the translational motion of its center of mass and the rotational motion about the long axis through its center of mass, we will apply WKET in the form of Equation (7.37) from time t_0 to t_1, remembering that the cylinder starts from rest (i.e., $v_0 = 0$ and $\omega_0 = 0$):

$$\underbrace{KE_{\text{cm.i}}}_{0} + \underbrace{KE_{\text{rot.i}}}_{0} + W_{\text{net}} + W_{\text{net.torque}} = KE_{\text{cm.f}} + KE_{\text{rot.f}}. \qquad (7.45)$$

Here we need to consider the possibility that each force could change the translational *and* rotational kinetic energy of the cylinder through its work and the work of its torque, respectively. First, let's explicitly write the work terms for each force and torque:

$$W_{F_g} + W_{\tau_{F_g}} + W_n + W_{\tau_n} + W_{F_T} + W_{\tau_{F_T}} + W_{f_s} + W_{\tau_{f_s}} = KE_{\text{cm.f}} + KE_{\text{rot.f}}. \qquad (7.46)$$

We can now examine each term on the left-hand side carefully:

- $W_{F_g} = 0$ because the gravitational force is perpendicular to the cylinder's displacement.
- $W_{\tau_{F_g}} = 0$ because the gravitational force is exerted at the axis of rotation. Therefore, its torque is zero about this axis.

- $W_n = 0$ because the force $\vec{n}$ is perpendicular to the cylinder's displacement.
- $W_{\tau_n} = 0$ because the line of action of $\vec{n}$ passes through the axis of rotation. Therefore, its torque is zero about this axis.
- W_{F_T}: $\vec{F}_T$ is a constant force and is in the same direction as the displacement, so its work can be found by:

$$W_{F_T} = \vec{F}_T \cdot \Delta \vec{x}_{cm} = F_T \Delta x_{cm} \cos 0 = F_T \Delta x_{cm} = 10 \text{ N} \cdot 1.5 \text{ m} = 15 \text{ J}. \quad (7.47)$$

- $W_{\tau_{F_T}}$: Since both $\vec{F}_T$ and its position vector from the axis of rotation are constant, $\vec{\tau}_{F_T}$ is also constant. Furthermore, $\vec{\tau}_{F_T}$ has the same direction as the cylinder's angular displacement $\Delta \vec{\theta}$, as can be seen from the right-hand rule. Thus, the work of its torque can be found by:

$$W_{\tau_{F_T}} = \vec{\tau}_{F_T} \cdot \Delta \vec{\theta} = \tau_{F_T} \Delta \theta \cos 0 = F_T R \sin \left(\frac{\pi}{2} \right) \Delta \theta = F_T R \Delta \theta. \quad (7.48)$$

Since there is no slipping, $R\Delta \theta = \Delta x_{cm}$, and we obtain from the previous equation:

$$W_{\tau_{F_T}} = F_T \Delta x_{cm} = 10 \text{ N} \cdot 1.5 \text{ m} = 15 \text{ J}. \quad (7.49)$$

The last two results show that the work done by $\vec{F}_T$ is the same as the work done by $\vec{\tau}_{F_T}$.

- W_{f_s}: $\vec{f}_s$ is also a constant force and is in the same direction as the displacement. Thus:

$$W_{f_s} = \vec{f}_s \cdot \Delta \vec{x}_{cm} = f_s \Delta x_{cm} \cos 0 = f_s \Delta x_{cm}. \quad (7.50)$$

- $W_{\tau_{f_s}}$: Since both $\vec{f}_s$ and its position vector from the axis of rotation are constant, $\vec{\tau}_{f_s}$ is also constant. However, $\vec{\tau}_{f_s}$ points in the opposite direction as the cylinder's angular displacement $\Delta \vec{\theta}$, as can be seen from the right-hand rule. Thus, the work of its torque can be found by (remembering that there is no slipping):

$$W_{\tau_{f_s}} = \vec{\tau}_{f_s} \cdot \Delta \vec{\theta} = \tau_{f_s} \Delta \theta \cos \pi = -f_s R \sin \left(\frac{\pi}{2} \right) \Delta \theta = -f_s R \Delta \theta = -f_s \Delta x_{cm}. \quad (7.51)$$

By adding the last two results together we see that the total work done by $\vec{f}_s$ (translational and rotational) is zero! This is a general result for rolling motion without slipping: $\vec{f}_s$ will either help or hinder the translational motion but it will have the opposite effect on the rotational motion, so that $W_{\text{total.fs}} = 0$.

Based on this discussion, and using $v_{cm} = \omega R$ (since there is no slipping), Equation (7.46) becomes:

$$W_{F_T} + W_{\tau F_T} = KE_{\text{cm.f}} + KE_{\text{rot.f}}$$

$$= \frac{1}{2}Mv_{\text{cm}}^2 + \frac{1}{2}I\omega^2$$

$$= \frac{1}{2}Mv_{\text{cm}}^2 + \frac{1}{2}\left(\frac{1}{2}MR^2\right)\omega^2$$

$$= \frac{1}{2}Mv_{\text{cm}}^2 + \frac{1}{4}Mv_{\text{cm}}^2 = \frac{3}{4}Mv_{\text{cm}}^2. \qquad (7.52)$$

We can solve this equation for v_{cm}, plug in the values for the two work terms and the cylinder's mass, and obtain the final velocity of the center of mass:

$$v_{\text{cm}} = \sqrt{\frac{4(W_{F_T} + W_{\tau F_T})}{3M}} = \sqrt{\frac{4 \cdot 30 \text{ J}}{3 \cdot 0.4 \text{ kg}}} = 10 \text{ m/s}. \qquad (7.53)$$

Exercises

(i) Find the magnitude and direction of the total acceleration of points A, B, and D in Figure 7.4 in terms of a_{cm} and a_c.

(ii) A uniform billiard ball is struck by a cue in such a way that the line of action of the cue's applied force is parallel to the table and passes through the center of the ball. The initial velocity v_0 of the ball after impact, its radius R, its mass m, the acceleration due to gravity g, and the coefficient of kinetic friction μ_k between the ball and the table are all known. For some time, the ball will be both slipping forward and rolling, but eventually, pure rolling will start. Find (a) how far the ball will move before it starts pure rolling and (b) the ball's angular velocity at this instant. The moment of inertia of a solid sphere (billiard ball) about an axis passing through its center is $I = \frac{2}{5}MR^2$.

(iii) Figure 7.11 shows a disk of radius R whose plane is perpendicular to a smooth horizontal surface. A massless, inelastic rope is wrapped around a groove a distance r from the disk's center and does not slip against the groove. A force $\vec{F}$, which forms an angle ϕ with the horizontal, is exerted at the free end of the rope (point A). The disk rotates due to this force about a horizontal axis that goes through its center O and is perpendicular to its plane. The disk is supported by a base and the base can slide along the smooth horizontal surface, allowing in this way the disk to both rotate and translate (but this is *not* rolling motion). At time t the disk has an angular velocity $\vec{\omega}$ and an angular acceleration $\vec{\alpha}$. Finally, the disk's center of mass has linear velocity $\vec{v}_{cm}$ and linear acceleration $\vec{a}_{cm}$. Just

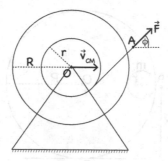

Figure 7.11 A disk, whose plane is perpendicular to a smooth horizontal surface, has a massless, inelastic rope wrapped around it at a groove a distance r from the disk's center O. The disk rotates due to a force $\vec{F}$ about a horizontal axis that goes through its center and is perpendicular to its plane. The force $\vec{F}$ is exerted at the free end of the rope (point A), and forms an angle ϕ with the horizontal. The disk is supported by a base that can slide along the horizontal surface.

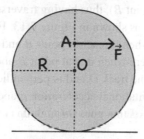

Figure 7.12 A disk on a rough horizontal surface with a force $\vec{F}$ applied at a point A on the disk that belongs to its vertical diameter. The position of point A along the vertical diameter determines the static friction $\vec{f_s}$.

like we did in Section 7.2.3, derive expressions for (a) the velocity of point A, (b) the acceleration of point A, (c) the displacement of point A in the direction along the rope, and (d) the length of the rope that gets unwound during a time dt.

(iv) A disk of mass M and radius R is initially at rest on a rough horizontal surface. At time $t_0 = 0$, we start exerting a force $\vec{F}$ that is always applied at a point A on the disk that belongs to its vertical diameter, as seen in Figure 7.12. Find the location of point A relative to the center O so that the static friction $\vec{f_s}$ equals zero. What happens to $\vec{f_s}$ if the force $\vec{F}$ is applied above or below that point A?

(v) A uniform sphere of mass m and radius r is released from rest from point A of a vertical, circular track of radius R and center O. The sphere rolls without slipping in the interior surface of the track and reaches the

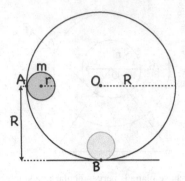

Figure 7.13 A small sphere of radius r and mass m is released from rest from point A of a circular track of radius R and rolls without slipping on the track's interior surface. When it reaches the bottom of the track (point B), the sphere has traversed a quarter of the circular track.

bottom of the track (point B), thus having traversed a quarter of the track's circumference, as shown in Figure 7.13. Find (a) the kinetic energy of the sphere when it reaches point B and (b) the angular momentum of the sphere when it is at point B about an axis that goes through the center of the track O and is perpendicular to the track's plane. You are given the gravitational acceleration g and the moment of inertia of a sphere about an axis that goes through its center of mass $I_{cm} = \frac{2}{5}mr^2$.

8

Garden of Delights

We have now reached the last chapter of this book! Therefore, it might appear to the reader that by the end of this chapter, we will have covered all the concepts on rotational motion and nothing else will be left to study on this topic.

Actually, this could not be further from the truth. In this book, as explained in the Preface and at the beginning of Chapter 1, our treatment was limited to rotation of *point masses* and *rigid bodies* about a *fixed* axis. We have not looked at nonrigid bodies or nonfixed axes of rotation, both of which give rise to more complicated but intriguing phenomena. Studying these in detail involves more advanced mathematics (e.g., matrix algebra), which is why these topics are studied in advanced mechanics courses.

However, we feel that this book would be doing a disservice to the reader if we did not point out how the concepts we have learned thus far can provide us with a fundamental understanding of some of these more advanced topics. Therefore, this last chapter is dedicated to just that: We would like to show how the toolkit and mastery of the concepts we have developed in these past seven chapters can facilitate our introduction to the more complicated phenomena. These phenomena will not be treated in as much detail as we have provided in our explanations thus far – we will only be providing some foundational knowledge. Nevertheless, by doing so we hope to further excite the students and readers of this book about what the future holds so that they look forward to continuing their study of mechanics and physics in general. There are still a lot of interesting problems ahead!

In this chapter, we will focus on three topics that are often found in upper-level mechanics courses. These are (a) gyroscopic precession, (b) calculations of angular momenta about points rather than axes, and (c) rotation matrices.

8.1 Gyroscopic Precession

The situation depicted in Figure 8.1 is as follows: A wheel has an axle that extends on both sides of the wheel's plane. At one end of the axle, point O, we tie a rope. The other end of the rope is firmly attached to the ceiling. If we let the wheel go from the initial position where the axle is horizontal at time t_i (Figure 8.1a), the wheel will rotate due to the gravitational force until the axle becomes vertical at time t_f (Figure 8.1b).

Why does the wheel rotate until the axle becomes vertical? This is quite simple to explain. There are two forces exerted on the wheel: The force of tension $\vec{F}_T$ and the gravitational force $\vec{F}_g$. Point O is a fixed point for the wheel and the wheel can rotate about the x axis. Because the force due to tension is exerted at O, it does not produce torque about the x axis. However, the gravitational force is exerted at a distance r from point O along the y axis at time t_i. Therefore, it produces a torque $\vec{\tau}_{F_g}$ along the x axis that causes the wheel to rotate about this axis. As the wheel rotates, $\vec{F}_g$ does not change direction, and since the angle between $\vec{F}_g$ and the position vector $\vec{r}$ changes, the torque decreases. When the axle is vertical at time t_f, the force $\vec{F}_g$ is parallel to the position vector, and thus no longer produces torque. Thus, in the end, the wheel remains at this final equilibrium position.

As a second experiment, let's spin the wheel about its axle before letting it go from its initial vertical position. If we set the wheel spinning, we will observe that the wheel no longer topples over when we release it, but remains

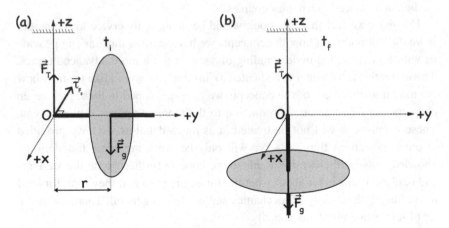

Figure 8.1 (a) A wheel with a long axle is attached to the ceiling via a rope connected at end O of the axle. Initially, the wheel is positioned so that the axle is horizontal and is released from rest at time t_i. (b) The gravitational force will cause the wheel to rotate about a horizontal axis that goes through point O (the x axis) until the axle becomes vertical at time t_f.

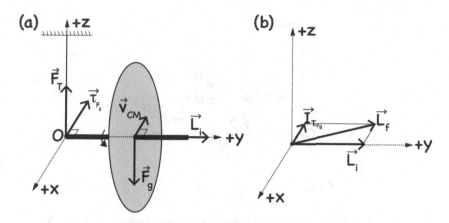

Figure 8.2 (a) The same wheel as in Figure 8.1 is now set to spin counterclockwise, as seen from the $+y$ axis, about its horizontal axle and then released. If it spins, the axle does not become vertical, but rather rotates in the xy plane (i.e., precesses) in the counterclockwise direction as seen from the $+z$ axis. (b) During each time interval dt, the torque due to $\vec{F}_g$ produces an impulse $\vec{\mathcal{I}}_{\tau_{Fg}}$. This impulse is always perpendicular to the instantaneous angular momentum. Thus, after each dt, the vector sum of the initial angular momentum $\vec{L}_i$ and the impulse will result in a final angular momentum vector $\vec{L}_f$ that precesses about the z axis.

vertical while at the same time rotating with a constant speed v_{cm} in the xy plane about an axis along the rope (the z axis), as shown in Figure 8.2a. The motion of the wheel now consists of two rotations: A fast rotation about the axle, which we call spin, and a slow rotation of the axle about the z axis, which we call precession. But why are there suddenly two axes of rotation? And why is the motion so different from when the wheel was not initially spinning?

One can find many videos of this weird (at first glance) phenomenon of the spinning wheel, known as gyroscopic precession. It is often done as a demonstration in introductory physics courses, which results in students' jaw-dropping, wide-eyed amazement. To explain this surprising phenomenon, first, let's remind ourselves of a simpler scenario from linear motion.

Imagine we have a small mass m moving along the y axis. If there are no forces exerted on the mass, it moves with constant momentum $\vec{p}_i$ along the y axis. Now, imagine that a force $\vec{F}$ is exerted on the mass that points in the $-x$ direction for a short duration dt, as shown in Figure 8.3. The force provides impulse $\vec{\mathcal{I}}_F$, which also points along the $-x$ axis. From MIT we know that

$$\vec{p}_i + \vec{\mathcal{I}}_{\text{ext}} = \vec{p}_f, \tag{8.1}$$

which means that the final momentum $\vec{p}_f$ after the interval dt is the vector sum of the initial momentum and the transferred momentum via the impulse

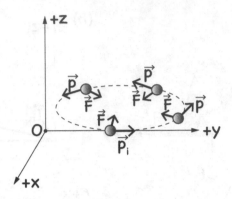

Figure 8.3 A mass is moving with constant momentum $\vec{p}_i$ along the y axis. At some time t, we exert a force $\vec{F}$ that points along the $-x$ axis for a short time dt. If the force continues to be exerted in a direction that is always perpendicular to the object's instantaneous linear momentum $\vec{p}$, then the object will execute circular motion since the force plays the role of the centripetal force.

$\vec{\mathcal{I}}_F = \vec{\mathcal{I}}_{\text{ext}}$. Because of this vector sum, the final momentum $\vec{p}_f$ will not be along the y axis but will veer off toward the x axis.

If we continue to exert this force in a direction perpendicular to the instantaneous momentum, we will see that the mass will execute circular motion, as our force plays the role of the centripetal force. Thus, the instantaneous momentum vector $\vec{p}$ continues "chasing" after the force vector, as shown in Figure 8.3. Because the force is always perpendicular to the instantaneous momentum, it is perpendicular to the instantaneous velocity and therefore the instantaneous displacement of the mass. Thus, it will not do work and will not change the magnitude of the momentum, just its direction.

Now let's apply our understanding of this simpler scenario to the spinning wheel. As is shown in Figure 8.2a, when the wheel is spinning counterclockwise (as seen from the $+y$ axis) about its axle, there is initial angular momentum $\vec{L}_i$, with direction along the axle (i.e., the $+y$ axis), as given by the right-hand rule. Now $\vec{L}_i$ is the equivalent of the linear momentum $\vec{p}_i$ just discussed. When we hang the wheel from the rope by attaching it at point O, we saw earlier that the gravitational force exerts a torque on the wheel about the x axis. Let's assume for now that this torque is exerted for an infinitesimal time interval dt. Of course, the torque is the equivalent to the force in the linear case discussed before. This torque is imparting rotational impulse $\vec{\mathcal{I}}_{\tau_{F_g}}$ to the wheel in the same direction as $\vec{\tau}_{F_g}$. From AMIT then we know that

$$\vec{L}_i + \vec{\mathcal{I}}_{\text{ext.r}} = \vec{L}_f, \tag{8.2}$$

which means that the final angular momentum $\vec{L}_f$ after the interval dt is the vector sum of the initial angular momentum $\vec{L}_i$ and the transferred momentum via the rotational impulse $\vec{\mathcal{I}}_{\tau_{F_g}} = \vec{\mathcal{I}}_{\text{ext.r}}$. Because of this vector sum, the final angular momentum $\vec{L}_f$, will not be along the y axis but it will veer off towards the x axis, as shown in Figure 8.2b.

Since this torque due to $\vec{F}_g$ will continue to be exerted in a direction perpendicular to the instantaneous angular momentum, the angular momentum vector of the wheel will continue to rotate in the xy plane. Thus, the instantaneous angular momentum vector continues "chasing" after the torque vector, just like in the linear case shown in Figure 8.3. Because the torque $\vec{\tau}_{F_g}$ is always perpendicular to the instantaneous angular momentum vector, it is perpendicular to the instantaneous angular velocity vector, and therefore the instantaneous angular displacement vector of the wheel. Thus, it will not do work and will not change the magnitude of the angular momentum, just its direction.

As we have been discussing the angular momentum of the wheel, we have only been talking about the spin angular momentum. This is the angular momentum we have indicated in Figure 8.2. However, as the wheel precesses about the rope, a second angular momentum is introduced! This is the orbital angular momentum of the center of mass of the wheel about the rope. In the example in Figure 8.2, this angular momentum would be along the $+z$ axis. This is our first encounter with a physical system in which there are multiple, simultaneous axes of rotation. As you advance through your studies, you will encounter rotations about multiple axes quite frequently, as this is the most general way to quantify rotations. Another new feature we have seen here is that our axis of rotation is no longer fixed – the spin axis of rotation changes its orientation in space as it precesses around the orbital axis of rotation (i.e., the z axis).

We have avoided a rigorous mathematical analysis here in favor of a more qualitative analysis so that we can explain the wheel's nonintuitive behavior. A more rigorous mathematical analysis shows that if the angular velocity of the precession $(\vec{\omega}_P)$ is much smaller in magnitude than the angular velocity of the wheel's spin $\vec{\omega}$ (i.e., $\omega_P \ll \omega$), then ω_P is inversely proportional to ω. If the condition $\omega_P \ll \omega$ is not satisfied, then the wheel will bob up and down (wobble) as it precesses. This wobbling is known as nutation.

The spinning wheel we have studied is not the only case of gyroscopic precession. For example, this phenomenon is also observed in a spinning top. For the top, the torque exerted by the gravitational force at its center of mass causes the spin axis to precess about an axis perpendicular to the table's surface.

8.2 Angular Momentum Revisited

As you know, the focus of all previous chapters of this book has been on rotation with respect to a fixed axis. Our discussion in the previous section on gyroscopic precession provided a first look at more general phenomena in rotational motion. In this section, we wish to generalize the idea of the angular momentum, which up until now has only been calculated with respect to an axis.

When we introduced the concept of angular momentum in Chapter 5, we defined it as $\vec{L} = \vec{r} \times \vec{p}$ for a point mass. In the scenarios we studied in the previous chapters, we were intentional about ensuring that $\vec{L}$ always pointed along the axis of rotation, in the same direction as the angular velocity $\vec{\omega}$. In general, this is not the case. Let's look at an example where $\vec{L}$ is not parallel to $\vec{\omega}$ and discuss the implications for a more advanced study.

We again consider the simple example of a point mass executing circular motion in the xy plane. The circular trajectory has radius r and its center is at the origin of the coordinate system O. The angular velocity $\vec{\omega}$ of the mass is along the z axis. At some time t, the mass is located on the y axis, as shown in Figure 8.4. We wish to find the angular momentum about the z axis, $\vec{L}_z$.

To find $\vec{L}_z$ we simply use the definition $\vec{L}_z = \vec{r} \times \vec{p}$, where $\vec{r}$ is the perpendicular distance between the mass and the z axis (in this case it is also the position vector of the mass with respect to the z axis) and $\vec{p}$ is the linear momentum.

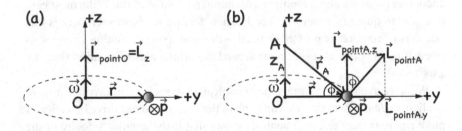

Figure 8.4 A point mass executes counterclockwise (as seen from above) circular motion on the xy plane. The center of the circular trajectory is at the origin O of the coordinate system and the angular velocity $\vec{\omega}$ of the mass is along the z axis. (a) At time t the mass is located on the y axis and has a linear momentum pointing into the page. The angular momentum of the mass about the z axis ($\vec{L}_z$) is equal to the angular momentum about the origin ($\vec{L}_{\text{pointO}}$) since the perpendicular distance between the z axis and the mass is the same as the distance between the origin and the mass. (b) The angular momentum of the mass about point A ($\vec{L}_{\text{pointA}}$) is not along the z axis but forms an angle ϕ with the z axis. However, for the z component of $\vec{L}_{\text{pointA}}$, we find that $\vec{L}_{\text{pointA.z}} = \vec{L}_{\text{pointO}} = \vec{L}_z$.

The right-hand rule gives that $\vec{L} = \vec{L}_z$ is along the z axis, as expected (Figure 8.4a). Furthermore, the magnitude is:

$$L_z = rp \sin\left(\frac{\pi}{2}\right) = rp. \tag{8.3}$$

At this point, we have not done anything unusual.

Now let's try something new. Instead of finding the angular momentum about the z axis, let's instead find it about the origin O. In other words, we now wish to find the angular momentum about a *point* (i.e., $\vec{L}_{\text{pointO}}$), not about an axis. The position vector of the mass about the origin is still $\vec{r}$, just like before, and the linear momentum is still $\vec{p}$. The right-hand rule again gives that $\vec{L}_{\text{pointO}}$ is along the z axis (Figure 8.4a). Furthermore, the magnitude is:

$$L_{\text{pointO}} = rp \sin\left(\frac{\pi}{2}\right) = rp = L_z. \tag{8.4}$$

In other words, $\vec{L}_z = \vec{L}_{\text{pointO}}$. In this case, it makes no difference whether we are evaluating the angular momentum with respect to the axis z or the point O, because the distance between the origin and the mass is the same as the perpendicular distance between the z axis and the mass. This is what we saw in all our examples in previous chapters.

Now, let's find the angular momentum of this mass at time t about a different point that is located on the z axis a distance z_A above the origin (point A). We will call this angular momentum L_{pointA}. In this case, the position vector of the mass with respect to point A is $\vec{r}_A$ but its linear momentum is still $\vec{p}$. The right-hand rule for the cross product $\vec{L}_{\text{pointA}} = \vec{r}_A \times \vec{p}$ gives that the direction of $\vec{L}_{\text{pointA}}$ is not along the z axis. As shown in Figure 8.4b, $\vec{L}_{\text{pointA}}$ is perpendicular to both $\vec{r}_A$ and $\vec{p}$, as expected from the definition of the cross product. Given that the angle between $\vec{r}_A$ and $\vec{p}$ is still $\frac{\pi}{2}$, the magnitude of $\vec{L}_{\text{pointA}}$ is given by:

$$L_{\text{pointA}} = r_A p \sin\left(\frac{\pi}{2}\right) = r_A p. \tag{8.5}$$

$\vec{L}_{\text{pointA}}$ forms an angle ϕ with respect to the z axis, which is the same angle as the one $\vec{r}_A$ forms with the y axis. Let's use this angle to find the component of $\vec{L}_{\text{pointA}}$ along the z axis, $\vec{L}_{\text{pointA.z}}$. The magnitude of this component is:

$$L_{\text{pointA.z}} = L_{\text{pointA}} \cos \phi = r_A p \cos \phi, \tag{8.6}$$

where we have used Equation (8.5). However, from the right triangle formed by r, z_A, and r_A we see that $r_A \cos \phi = r$. Therefore, Equation (8.6) becomes

$$L_{\text{pointA.z}} = rp, \tag{8.7}$$

which is the same as our results for L_z and L_{pointO} in Equations (8.3) and (8.4), respectively! Therefore, $\vec{L}_{pointA.z} = \vec{L}_z = \vec{L}_{pointO}$. Of course, as can be seen in Figure 8.4b, there is also a component of $\vec{L}_{pointA}$ along the y axis, $\vec{L}_{pointA.y}$ which will be equal to:

$$L_{pointA.y} = L_{pointA} \sin \phi = r_A p \sin \phi = z_A p, \qquad (8.8)$$

where we have used that $r_A \sin \phi = z_A$. In addition, we see that there is no component of $\vec{L}_{pointA}$ along the x axis ($\vec{L}_{pointA.x} = 0$).

We could have just as easily found the x, y, and z components for $\vec{L}_{point\,A}$ by calculating the cross product. Given that $\vec{r}_A = r\hat{j} + z_A(-\hat{k})$ and $\vec{p} = p(-\hat{i})$ we get:

$$\begin{aligned} \vec{L}_{pointA} &= \left(r\hat{j} + z_A\left(-\hat{k}\right)\right) \times p\left(-\hat{i}\right) \\ &= rp\left(\hat{j} \times \left(-\hat{i}\right)\right) + z_A p\left(\left(-\hat{k}\right) \times \left(-\hat{i}\right)\right) \\ &= rp\hat{k} + z_A p\hat{j}, \end{aligned} \qquad (8.9)$$

which agrees with our results from Equations (8.7) and (8.8) for the components of $\vec{L}_{pointA}$.

So what is the significance of these results? First, we see that angular momentum can also be found relative to a point, not just an axis. Nevertheless, regardless of the reference point, the component of the angular momentum along the z axis is the same in all cases. What this means is that when we find the angular momentum about an axis, what we are really finding is the *projection* of the angular momentum vector on that axis. Therefore, finding the angular momentum about a point is the most general case.

In addition, we see that depending on where we pick our reference point about which we find the angular momentum, $\vec{L}$ does not necessarily point in the same direction as $\vec{\omega}$. Thus, in the most general case, we can no longer use $\vec{L} = I\vec{\omega}$ in the usual form. The special set of axes for rigid body rotation in which $\vec{L}$ and $\vec{\omega}$ are parallel are called the principal axes of rotation. In this case, we can apply $\vec{L} = I\vec{\omega}$ about each principal axis separately, where I is a principal moment of inertia. In cases where we don't know the principal axes, we will need to compute the moment of inertia *tensor* to study rigid body rotations, which is a topic for another day!

8.3 Rotation Matrices

For our last topic in this chapter, let's introduce a new mathematical formalism for describing rotations which has its basis in linear algebra. If you are unfamiliar with linear algebra, don't get discouraged – hopefully this section can help you in the future, if not now. Because it is more mathematically advanced, we left this as the last topic in this book.

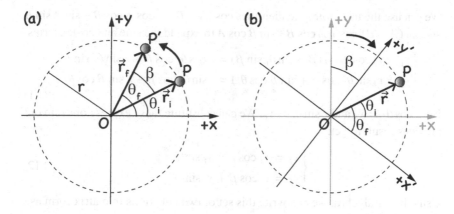

Figure 8.5 (a) A mass with initial position $\vec{r}_i$ rotates counterclockwise by an angle β about the z axis in the xy plane of a fixed Cartesian coordinate system until it reaches its final position $\vec{r}_f$. (b) The counterclockwise rotation of the mass about the z axis of the fixed coordinate system is equivalent to the clockwise rotation of the coordinate system by an angle β about the z axis while the mass remains stationary.

We again consider (for the last time!) the simple example of a point mass executing circular motion in the xy plane. The circular trajectory has radius r and its center is at the origin of the coordinate system O. Let's assume that the mass is moving in the counterclockwise direction as shown in Figure 8.5a. At any time t, the position of the mass is given by the Cartesian coordinates (x, y). The position of the mass can also be expressed using the polar coordinates (r, θ). Of course, we can easily go from polar to Cartesian coordinates since $x = r\cos\theta$ and $y = r\sin\theta$. We can also go from Cartesian to polar coordinates since $r^2 = x^2 + y^2$ and $\theta = \tan^{-1}\frac{y}{x}$.

As shown in Figure 8.5a, at time t_i the mass is at point P of its circular trajectory. The position vector $\vec{r}_i$ ($r_i = r$) of the mass forms an angle θ_i with respect to the $+x$ axis. In terms of its coordinates, it is at location $(x_i, y_i) = (r\cos\theta_i, r\sin\theta_i)$. Then, within the time interval Δt, let's assume that the mass has rotated by an angle β to reach the final position P'. What are the Cartesian and polar coordinates of the mass at time t_f?

In polar coordinates, the final position vector $\vec{r}_f$ ($r_f = r$) forms an angle of $\theta_f = \theta_i + \beta$ with respect to the $+x$ axis and thus the mass is located at $(r, \theta_i + \beta)$. In terms of the Cartesian coordinates (x_f, y_f) we have:

$$\begin{cases} x_f = r\cos\theta_f = r\cos(\theta_i + \beta) \\ y_f = r\sin\theta_f = r\sin(\theta_i + \beta). \end{cases} \tag{8.10}$$

We can use the trigonometric identities $\cos(A+B) = \cos A \cos B - \sin A \sin B$ and $\sin(A+B) = \sin A \cos B + \sin B \cos A$ to expand the sine and cosine terms:

$$\begin{cases} x_f = r(\cos\theta_i \cos\beta - \sin\theta_i \sin\beta) = r\cos\theta_i \cos\beta - r\sin\theta_i \sin\beta \\ y_f = r(\sin\theta_i \cos\beta + \sin\beta \cos\theta_i) = r\sin\theta_i \cos\beta + r\sin\beta \cos\theta_i. \end{cases}$$

$$(8.11)$$

But $r\cos\theta_i = x_i$ and $r\sin\theta_i = y_i$. We can substitute these expressions into our previous result to get:

$$\begin{cases} x_f = x_i \cos\beta - y_i \sin\beta \\ y_f = y_i \cos\beta + x_i \sin\beta. \end{cases}$$

$$(8.12)$$

Using linear algebra, we can write this set of two equations in matrix form as

$$\begin{bmatrix} x_f \\ y_f \end{bmatrix} = \begin{bmatrix} \cos\beta & -\sin\beta \\ \sin\beta & \cos\beta \end{bmatrix} \begin{bmatrix} x_i \\ y_i \end{bmatrix} \Rightarrow \vec{r}_f = R(\beta)\vec{r}_i, \qquad (8.13)$$

where we have defined $R(\beta) = \begin{bmatrix} \cos\beta & -\sin\beta \\ \sin\beta & \cos\beta \end{bmatrix}$.

Here we have established that a counterclockwise rotation in the xy plane can be represented via the matrix $R(\beta)$, which is called a rotation matrix. Given the initial position $\vec{r}_i$ of a mass and the angle β by which the mass rotates in the counterclockwise direction, its final position $\vec{r}_f$ can easily be found by the matrix multiplication of the rotation matrix $R(\beta)$ with $\vec{r}_i$.

If the rotation is in the clockwise direction, then we can simply substitute $-\beta$ for β in the rotation matrix. By making this substitution, we obtain:

$$R'(\beta) = R(-\beta) = \begin{bmatrix} \cos(-\beta) & -\sin(-\beta) \\ \sin(-\beta) & \cos(-\beta) \end{bmatrix} = \begin{bmatrix} \cos\beta & \sin\beta \\ -\sin\beta & \cos\beta \end{bmatrix}. \qquad (8.14)$$

But, you may recognize that $R'(\beta)$ is simply the transpose of R! In other words $R'(\beta) = R^T(\beta)$. What does this mean physically? This relationship shows that the final coordinates of a mass rotating counterclockwise in the xy plane with respect to a fixed coordinate system are equal to the coordinates obtained if the coordinate system rotates clockwise about the z axis while the mass remains fixed (Figure 8.5b). In other words, the rotation matrix $R(\beta)$ can be used to represent the counterclockwise rotation of the mass (or equivalently the clockwise rotation of the coordinate system), while the rotation matrix $R^T(\beta)$ can be used to represent the clockwise rotation of the mass (or equivalently the counterclockwise rotation of the coordinate system).

One can go through similar, but algebraically more cumbersome, derivations to find the rotation matrices in three dimensions of basic rotations, that is, rotations about the x, y, or z axis. The rotations matrices $R_x(\beta)$, $R_y(\beta)$, and $R_z(\beta)$ (for rotations about the x, y, and z axes, respectively) are:

$$R_x(\beta) = \begin{bmatrix} 1 & 0 & 0 \\ 0 & \cos\beta & -\sin\beta \\ 0 & \sin\beta & \cos\beta \end{bmatrix}, \quad R_y(\beta) = \begin{bmatrix} \cos\beta & 0 & \sin\beta \\ 0 & 1 & 0 \\ -\sin\beta & 0 & \cos\beta \end{bmatrix},$$

$$R_z(\beta) = \begin{bmatrix} \cos\beta & -\sin\beta & 0 \\ \sin\beta & \cos\beta & 0 \\ 0 & 0 & 1 \end{bmatrix}.$$

(8.15)

As an example in 3D, let's say we know the initial position $\vec{r}_i = (x_i, y_i, z_i)$ of a particle which rotates about the y axis by some angle β. We can find the final coordinates $\vec{r}_f = (x_f, y_f, z_f)$ by doing the multiplication $\vec{r}_f = R_y(\beta)\vec{r}_i$.

Rotation matrices also allow us to easily find the final position of a mass after it has undergone several consecutive rotations about different axes. For example, imagine that a mass with initial position $\vec{r}_i = (x_i, y_i, z_i)$ first rotates about the x axis by an angle θ_1 and then about the y axis by an angle θ_2. To find its final position $\vec{r}_f = (x_f, y_f, z_f)$ we first need to conduct the multiplication $R_x(\theta_1)\vec{r}_i$ to find the coordinates after the rotation about the x axis. We then multiply the result of this multiplication with $R_y(\theta_2)$. In other words $\vec{r}_f = R_y(\theta_2)(R_x(\theta_1)\vec{r}_i) = R_y(\theta_2)R_x(\theta_1)\vec{r}_i$. You might recall from linear algebra that the commutative property does not always hold for matrix multiplication and, as we discussed in Chapter 1, the commutative property does not hold for rotations in general either. In other words, though the product $R_y(\theta_2)R_x(\theta_1)\vec{r}_i$ will give the final position after the mass has rotated about the x axis first and then about the y axis, the product $R_x(\theta_1)R_y(\theta_2)\vec{r}_i$ will result in the final position after the mass has rotated about the y axis first and then about the x axis, with $R_y(\theta_2)R_x(\theta_1)\vec{r}_i \neq R_x(\theta_1)R_y(\theta_2)\vec{r}_i$. To summarize, the order in which we multiply the rotation matrices matters.

There are many more things we can say about rotation matrices by exploiting various properties of matrices in combination with theorems from linear algebra. Students of upper-level mechanics courses will spend a lot of time working with this formalism, especially after having a course on linear algebra. For now, we hope that this last chapter has whet your appetite to continue your study of physics at higher levels.

Further Reading

Carnero, C., Aguiar, J., and Hierrezuelo, J. 1993. The work of the frictional force in rolling motion. *Physics Education*, **28**(4), 225–227.

Goldstein, Herbert, Poole, Charles, and Safko, John. 2002. *Classical Mechanics*. 3 edn. Addison-Wesley.

Kleppner, Daniel, and Kolenkow, Robert. 2014. *An Introduction to Mechanics*. 2 edn. Cambridge University Press.

Mahajan, Sanjoy. 2020. *A Student's Guide to Newton's Laws of Motion*. Cambridge University Press.

Morin, David. 2013. *Classical Mechanics with Problems and Solutions*. Cambridge University Press.

Mungan, Carl E. 2001. Acceleration of a pulled spool. *The Physics Teacher*, **39**(8), 481.

Singh, Jitender, and Chaturvedi, Shraddhesh. 2019. *300 Solved Problems on Rotational Mechanics*. 1 edn. PsiPhiETC.

Streib, John F. 1989. A theorem on moments of inertia. *American Journal of Physics*, **57**(2), 181.

Taylor, John R. 2005. *Classical Mechanics*. University Science Books.

Πενέσης, Θοδωρής, and Συνοδινός, Διονύσης. 2014. *Φυσική Γ' Λυκείου Θετικής και Τεχνολογικής Κατεύθυνσης, Β' Τόμος*. 2 edn. Ελληνοεκδοτική.

Index

Printed in the United States
by Baker & Taylor Publisher Services